ALLE·ZEIT·WACH
1842

T. Sato · T. Iizuka (Eds.)

Color Atlas of Surgical Anatomy for Esophageal Cancer

With 105 Illustrations, Mostly in Color

Springer-Verlag
Tokyo Berlin Heidelberg New York London Paris
Hong Kong Barcelona Budapest

Professor TATSUO SATO, M.D.
2nd Department of Anatomy, School of Medicine, Tokyo Medical and Dental University, 1-5-45 Yushima, Bunkyo-ku, Tokyo, 113 Japan

Professor TOSHIFUMI IIZUKA, M.D.
National Oji Hospital, 4-17-56 Akabane-dai, Kita-ku, Tokyo, 115 Japan

On the front cover: Lymph nodes along the right margin of the trachea and esophagus in the thorax; *see p. 31*.

ISBN 978-4-431-68200-4 ISBN 978-4-431-68198-4 (eBook)
DOI 10.1007/978-4-431-68198-4

Printed on acid-free paper.

Library of Congress Cataloging-in-Publication Data
Color atlas of surgical anatomy for esophageal cancer / T. Sato, T. Iizuka (eds.). p. cm.
Includes bibliographical references and index. ISBN 978-4-431-68200-4
1. Esophagus—Cancer—Surgery—Atlases. 2. Lymph nodes—Anatomy—Atlases. 3. Esophagus—Anatomy—Atlases. 4. Anatomy, Surgical and topographical—Atlases. I. Sato, Tatsuo, 1937– . II. Iizuka, T. (Toshifumi, 1928– . [DNLM: 1. Esphageal Neoplasms—surgery—atlases. WI 17 C7192] RD667.3.C65 1992. 616.99'432059—dc20. DNLM/DLC. for Library of Congress. 92-49824

Softcover repint of the hardcover 1st edition 1992

Foreword

Without knowledge of anatomy a surgeon is like a "mole". This verdict of Sauerbruch has kept its importance until today. Moreover, this importance is growing with the trend in modern cancer surgery to remove not only the involved organ but also the adjacent structures around the organ, particularly the lymph pathways. Modern lymph node dissection requires exact anatomical and embryogenetic information.

Lymph node dissection means having to trace back the embryogenic development of the organ, here the esophagus. Because the esophagus has two different embryogenetic origins—branchial arches and yolk sac—which grow together during embryogenetic development in the area of the tracheal bifurcation—the organ also has two main lymphatic collection areas, in the abdomen and at the neck. The yolk sac has its main lymphatic collection around the celiac trunk. There are only few lymph nodes around the distal esophagus. The upper lymphatic collection is much more difficult because the trachea and esophagus have the same embryogenetic origin (branchial arches); separation takes place late in embryogenetic development. All the lymphatic pathways are located on both sides of the trachea and in the area of the cervical vessels.

In any location the lymphatic pathways are very close to the blood supply of the organ. The same is true for the esophagus. The blood supply comes from below (celiac trunk, left gastric artery, etc.) and from above (lower thyriod artery, etc.). There are only very few direct branches from the aorta to the esophagus in the area of the bifurcation.

It is the great merit of the Japanese School of Surgery that they have described the correct method of lymph node dissection, eradicating all three collecting areas of the lymph pathways: in the area of the two embryogenetic origins (celiac trunk and neck), as well as the connecting area of both esophageal origins in the mediastinum—a procedure correctly called three-field lymph node dissection. According to the localization of the primary tumor, lymph node dissection of the upper two fields or the lower two fields is carried out, while in a tumor localized at the bifurcation all three fields must be surgically cleared.

This "Color Atlas of Surgical Anatomy for Esophageal Cancer" gives the surgeon who wants to perform adequate esophageal surgery the optimal anatomical and surgical fundamentals to do this. It is a gift to all esophageal surgeons. Because this book was written under the auspices of the "ISDE" it is something like the "ISDE Textbook of Modern Esophageal Surgery".

Hoping it will have a wide distribution.

J. Rüdiger Siewert

Foreword

Surgery for esophageal diseases is among the most difficult for any organ. The location of the esophagus, crossing through three body regions, cervical, thorax, and abdomen, and in close proximity to vital organs, makes surgical resection of the esophagus a formidable operation. Cancer of the esophagus has a well-deserved reputation for being deadly, in part because of the anatomical site of the organ, and in part because there are few, if any, early symptoms before the entire esophageal circumference is involved, causing dysphagia. Surgery for esophageal cancer is now in its sixth decade of common practice since being introduced by Ohsawa in Japan, as reported in 1933. For nearly five decades, cure from resection was regarded as a random and unpredictable event. Indeed, some propose that esophageal cancer is a systemic disease from the onset because the survival rates are so poor. Death from esophageal cancer generally occurs rapidly, and those succumbing to the disease are nearly all dead in less than 3 years. Those who survive a longer time are often cured and die of natural causes. Study of such survivors provides a basis for developing a curative strategy.

The necessity to identify candidates for surgical cure stems from the introduction of *en bloc* resection, in which an effort is made to achieve complete local and regional control around the primary cancer. Some patients with spread into the regional lymph nodes and adjacent tissues are cured by this more extensive procedure. In fact, the first patient successfully treated by one-stage esophagectomy and esophagogastrostomy in the Western world, reported by Adams and Phemister in 1938, survived more than 15 years in spite of a positive lymph node in the resected specimen.

In the 1990s, there are several competing techniques striving for improved cure rates of esophageal cancer. These include several variations of extensive surgery, and combinations of therapy, including chemotherapy and irradiation. To enable comparison of results from different treatments, precise staging of the disease is essential. Without knowing the extent of the disease, it is not possible to decide which treatment may be more beneficial. For other cancers of the digestive tract, such as stomach and colon cancer, it is long accepted that attempted curative surgery must include resection of the regional lymph nodes. However, the difficulties in defining the region to be resected for esophageal cancer and the lack of general understanding of the lymphatic anatomy and drainage pathways make the design and comparison of surgical techniques much more controversial and difficult than for treatment of cancer in other digestive tract organs. With extensive data now available internationally which indicate that patients with lymph node metastases can be cured by extensive surgery, controversies about the extent of surgery are among the major debates at this time. To resolve these questions, not only the number of nodes dissected, and the number of positive nodes, but also the location of the lymph nodes in relation to the primary tumor must be recorded. Only precise identification of the location and numbers of lymph nodes will permit resolution among the proponents of several types of esophagectomy, including three-field lymph node dissection, two-field dissection, *en bloc* resection, and variations on a theme of esophagectomy without thoracotomy.

This Atlas is of enormous benefit to students of esophageal cancer as it provides a clear description of the anatomical features of esophageal surgery. The lymphatic pathways are clearly illustrated. The current staging system and controversies in staging are described. A basis for international understanding of the staging system is well presented by this book.

The Japanese authors of this Atlas are pioneers in modern esophageal surgery. They have achieved an admirable national collaboration with colleagues to develop a major data bank on esophageal cancer which can be correlated with survival rates. These data serve as a factual basis for this Atlas. Each author contributing to this book is internationally known, and each is making a major contribution to the subject of surgery for esophageal cancer. They are ideally qualified to present this volume of major international importance, a volume which will set a basis for better understanding and resolution of controversies in this field.

David B. Skinner

Preface

It is essential to know the extent of lymph node metastasis when we operate for esophageal carcinoma, because metastases to the distant lymph nodes are frequently observed even in cases of superficial lesions of the esophagus. Metastasis can extend not only to the mediastinal but also to the neck and abdominal lymph nodes. So, three-field dissection of the lymph nodes has been performed for thoracic esophageal carcinoma in Japan for the last 10 years. However, the rationale for extended three-field dissection has not been established in the past. Many Japanese surgeons have been studying the incidence of metastasis to each lymph node group, and therefore, appropriate maps of lymph node groups are required.

This volume contains five papers concerning the dissection of lymph nodes for carcinoma of the thoracic esophagus. Professor Sato presents maps of the lymph nodes of the mediastinum and neck in colourful schematic illustrations, and classifies the lymph nodes into groups. He also shows the lymphatic pathways connecting these lymph nodes based on careful dissection of human corpses. These illustrations will be useful for surgeons performing reasonable operations for esophageal carcinoma. His work was carried out at the request of the TNM Research Committee of the International Society for Diseases of the Esophagus. Professor Akiyama presents the background of lymph node dissection for squamous cell carcinoma of the esophagus. He stresses the necessity for extended lymph node dissection, which will lead to improvement of survival after resective surgery. Professor Kakegawa presents illustrations of surgery for carcinoma of the thoracic esophagus, and methods for accurately recording the dissected lymph nodes. This may be useful for a comparison of the incidence of metastasis to lymph nodes between different institutions in foreign countries.

Through this collection of papers, readers will gain an understanding of the methods of reasonable dissection of the lymph nodes for carcinoma of the thoracic esophagus based on the lymph node classification made by Professor Sato. Careful dissection of lymph nodes may be the best way to improve survival rates after surgery for cancer of the thoracic esophagus.

Finally, we should like to express our hearty thanks to the Japanese Foundation for Multidisciplinary Treatment of Cancer for its generous aid with this publication.

Tatsuo Sato
Toshifumi Iizuka

Preface

Contents

List of Contributors

Milestones Along the Road to Improvement of Results in the Treatment of Squamous Cell Carcinoma of the Esophagus

KIYOSHI INOKUCHI[1]

Introduction

A diagnosis of cancer of the esophagus was tantamount to a death sentence until 40 years ago, at which time the treatment of this disease held out virtually no hope for the patient. However, the prognosis after surgical treatment has been improved considerably. Following the work of Torek [1] in 1913, who reported the successful resection of thoracic esophageal cancer, and later Ohsawa [2] in 1933, who demonstrated esophagectomy to be a practical surgical concept, surgery for this disease was developed by Adams and Phemister [3] in 1938. Later, Garlock [4] and Sweet [5] confirmed this type of operation to be the standard procedure for thoracic esophageal cancer.

Despite such heroic efforts, the operative mortality was over fifty percent. Nakayama strove to decrease the operative mortality, and aiming to avoid the fatal risk of incidental leakage at the site of intrathoracic anastomosis, in 1948 he adopted the antesternal route for esophagectomy [6], which had first been reported by Kirschner. Nakayama [7] also proposed a multi-staged operation to increase the safety of the procedure. Furthermore, Nakayama et al. [8] studied the efficacy of adjuvant radiotherapy on squamous cell carcinoma and the concept of preoperative radiotherapy prevailed thereafter. Such efforts in Japan in the 1960s led to a considerable improvement in operative mortality, to approximately ten percent.

As to the current status of the results of treatment of esophageal cancer, when Earlam and Cunha-Melo [9] in 1980 reviewed over 83 000 patients, they found that the overall rate of resection was 39% and that 18% of patients with successful resections survived for 5 years. The operative mortality reported varied widely, ranging from 0.8% to 57%. Giuli and Gignoux [10] reported 30% operative mortality in a review of 2400 cases. There was therefore tremendous variation in the results of surgery.

The Japanese Committee of Registration of Esophageal Cancer [11] reviewed 7742 patients from 119 hospitals in total; the operative mortality was 7.3%. Of those who were discharged, 22.5% survived for 5 years.

[1] Saga Prefectural Hospital, Mizugae, Saga, 840 Japan

Subsequent review [12] was made of 5024 patients from 141 hospitals between 1979 and 1982. The figures were 67.5% for resectability, 6.8% for operative mortality, and a 26.08% 5-year survival rate. Gradual improvement of the results of surgical treatment was shown. Recently, the concept of extensive surgery has become prevalent in Japan.

The data reported herein vary greatly. Some investigators in countries reporting high mortalities may mistakenly think that such variations might be due to some difference in the nature of the malignancy of the disease. However, the broad spectrum of figures is probably produced on the one hand by conservative surgeons, who do not want to subject a patient to an operation which carries the risk of death, and on the other hand by aggressive surgeons with varying levels of expertise, who are attempting to treat as many patients as possible. The stage at diagnosis, expertise of the surgeons, patient selection, radicality of operation, and pre- and postoperative care, among other factors, are all variable parameters.

The abundance of the literature on this subject presents a chaotic aspect. We must select the data of real value for future studies. In order to convert such a broad spectrum of figures to a comprehensible level, the data must be compiled on the basis of a uniform scale. Appropriate hypotheses, based on such information and on controlled randomized prospective studies, must be logically established; such hypotheses are necessary to further clarify the effective management of therapeutic strategies for cancer of the esphagus.

Appraisal of Extended Surgery for Esophageal Cancer

It is a general principle to dissect regional lymph nodes as much as possible, although the value of extensive lymph node dissection is not always unanimously agreed upon in esophageal cancer. The train of my thoughts reminds one of the Japanese work on gastric cancer surgery, in which a general policy of prophylactic lymph node dissection during surgery, adopted since the 1970s, has strikingly elevated the long-term survival rate to more than twice that previously obtained by simple gastrectomy.

It is well known that the presence of lymph node metastasis in esophageal cancer is the most important factor contributing to a poor prognosis. Endo et al. [13] reported that the 5-year survival rate of superficial cancer with no lymph node metastasis was 62%, whereas in cases of positive node metastasis the figure dropped to a mere 17%, which is almost the same as that of advanced cases. A Japanese national survey of superficial esophageal cancer patients treated from 1969 to 1978 [11] reported the 5-year survival rate of patients without nodal metastasis to be 60.8%, while that of those with metastasis was 30.8%. It is thus apparent that the presence or absence of lymph node metastasis significantly affects the prognosis of esophageal cancer.

Analysis of lymph nodes in cases of esophageal cancer revealed widespread distribution of positive lymph node metastasis extending from the superior mediastinum to the abdominal celiac region, regardless of the location of the primary tumor [14]. Surgeons dealing with extensive node dissection perform surgery in both the thoracic and the abdominal areas, and sometimes in the cervical area, to dissect supraclavicular as well as deep cervical nodes.

Table 1. Five-year survival rate of extensive surgery for thoracic esophageal cancer (curative cases).

Author	*Number of cases*	*Operative mortality*	*Five-year survival rate*		
			Total	*Negative metastasis*	*Positive metastasis*
Akiyama [15]	205	1.4%	34.6%[a]	53.8%[a]	15.3%
Nishihira [16]	408	11.3%	36.0%[a]	65.1%[a]	17.8%
Kakegawa [17]	116	4.3%	34.9%[b]	57.8%[a]	19.4%
Watanabe [18]	122		39.0%[a]	55.0%[a]	16.0%
National Survey, Japan [12]	3170	6.94%	25.99%	39.40%	17.30%
			[a]$P < 0.01$ [b]$P < 0.05$	[a]$P < 0.01$	n.s.

[a], [b] are statistically significant, on the χ^2 test, as compared to those of a National Survey in Japan. (last line)
n.s., not significant

Table 2. Five-year survival rate following simple and extensive gastrectomy for patients with gastric cancer (Kodama [14]).

	Number of cases	*Total*	*Negative metastasis*	*Positive metastasis*
Simple gastrectomy	254	33%	73%	18%
Extended gastrectomy	454	58%	81%	39%
		$P < 0.001$	n.s.	$P < 0.001$

Table 1 reviews the clinical results of extensive surgical procedures in Japan [15–18]. In this Table, the figures from a Japanese National Survey from 1979 to 1982 are referred to as conventional transthoracic surgery, which involved a general policy of dissecting peri-esophageal and some para-esophageal lymph nodes. In the extended surgery series, the operative mortality ranged from 1.4% to 11.3%, showing that extended surgery could be performed while keeping the mortality within an acceptable level.

The 5-year survival rate of extensive surgery (curative cases) ranged from 34% to 39%, significantly higher than that of the conventional series, 26%. Extensive surgery procedures were significantly effective for patients without lymph node metastasis, while in those with nodal metastasis the results were the same as those of conventional surgery.

These figures are different from those for cases of gastric cancer surgery, as shown in Table 2, which indicates representative data reported from Japan [14]. Extensive surgery was significantly effective only in patients with lymph node metastasis; it was not effective in patients without node metastasis.

This characteristic difference may reflect some of the biological features of esophageal cancer. One possible explanation may be that esophageal cancer has a greater tendency to spread to more distant areas than gastric cancer, and that the lesions with nodal metastasis involve a wider spread beyond the surgically resectable range, so that even extensive surgery might prove useless. However, most lymph nodes judged to be negative for metastasis on routine examination actually contain small or occult micrometastatic foci, which can be resected by surgery. Since such micrometastatic foci spread within a certain range acceptable by extensive node dissection, it may be that extensive surgery will prove to be an effective treatment only for patients without node metastasis.

Such paradoxical facets of the biological characteristics of esophageal cancer remind surgeons of controversies centered on criticism of extensive surgery. It may be common sense that extensive surgery has merit in advanced cancer cases. This approach has in fact been proven valid in gastric cancer; however, esophageal cancer appears to be different from gastic cancer in that extensive surgery is rather effective in less advanced case.

Orringer [19], one of those surgeons who does not support extensive surgery, has expressed the view that despite improvements in surgery, transthoracic esophagectomy and reconstruction is a formidable operation. He proposed the policy of transhiatal esophagectomy without thoracotomy. In his report of 100 patients with esophageal cancer, operative deaths occurred in 6 and the 5-year survival rate was 17%. He stressed that these figures are as good as those reported in many series of transthoracic esophagectomies. However, as several authors [20] have pointed out, it is difficult to accept the finding that transhiatal approaches can be performed with lower mortality and morbidity than transthoracic standard procedures. The former apparently has the limitation that lymph node dissection in areas adjacent to the lesion is not possible. However, an extrathoracic approach may be acceptable as an alternative in the surgical armamentarium.

In order to settle such controversies, prospective controlled trials are essential. For this purpose, it is indispensable to establish uniform international standards. The uniform staging of the disease determined by the revised TNM Classification should be widely used.

Combination Therapy

Considering the fact that extensive surgery might be less effective for patients with lymph node metastasis, we must look toward multidisciplinary treatment such as radio-chemo-immunotherapy. Although surgery and radiotherapy are both local treatments, chemotherapy differs in that it is generally a systemic therapy and might be expected to hold more hope of cure. However, the exact role of chemotherapy has not yet been settled, although occasional remission has been reported. In particular, reports on the effectiveness of chemotherapy in esophageal cancer patients with lymph node metastasis are almost nonexistent.

Nishihira et al. [21] carried out a systematic study to determine whether or not aggressive postoperative chemotherapy increased survival in patients with lymph node metastasis. Separate protocols for respective patient groups with regional lymph node metastasis (Protocol I) and with distant node metastasis (Protocol II) were prepared. Protocol I consisted of consecutive administration of Tegafur, at an oral dose of 600 mg/day or via a suppository of 750 mg/day and intravenous pepleomycin (100 mg in total) at a dose of 5 mg/day on the day of irradiation. Biological response modifiers, such as OK-432 (a streptococcus preparation) and/or PSK (a basidiomycetes preparation) were given immediately after radiotherapy and continued after discharge for as long as possible. The results of 67 patients with node metastasis n1(+)n2 (+) showed the 5-year survival to be 32.5%. Since the

5-year survival figures in a nationwide survey in Japan from 1974 to 1978 were 39.9%, 18.7%, 16.7%, and 6.05%, respectively, for patients with n(−), n1(+), n2(+) and n3(+), Protocol I is estimated to have recorded a significantly superior figure compared to the historical controls.

Protocol II, more aggressive than the Protocol I, was given to 19 patients with distant node metastasis. The regimen consisted of pepleomycin, adriamycin, and mitomycin, followed by cisplatin and vindesine, the dosage of which these agents was approximately twice that of the conventional regimen. The 3- and 5-year survival rates of patients were 30.5% and 19.0%, respectively, while no patients not receiving such treatment survived for more than three years. In this series, only 15% of patients with distant metastasis receiving Protocol I survived for two years, a significantly lower result than that for Protocol II. Considerable side effects, such as nausea and anorexia, occurred and the application of enteral or parenteral nutritional support was necessary to ensure complete administration of the regimen.

A final evaluation of this study is difficult, since it was not a prospective controlled randomized trial. However, considering that such figures as 32.5% 5-year survival in patients with regional node metastasis and 30.5% 3-year survival in patients with distant node metastasis have not been obtained by other studies, this study therefore strongly suggests the effectiveness of chemotherapy in advanced esophageal cancer. More analytical standard trials are anticipated.

Concept of Type-Oriented Therapy for Esophageal Cancer

As mentioned above, patients with positive regional lymph node metastasis have very poor survival despite extensive node dissection, but aggressive combined adjuvant therapy appears to prolong survival. On the other hand, for those patients with no metastasis, extensive surgery is effective in improving prognosis. It is suggested that lesions with no node metastasis behave as a local disease, whereas the ones with positive metastasis to regional lymph nodes are no longer local, but behave as a systemic disease. This idea must be reflected by the type of therapy. It is not wise to apply a uniform schedule of therapy to all patients regardless of different malignant potential. Aggressive combined treatment involves considerably sophisticated elements and is often accompanied by serious side effects, making us hesitate to apply this strategy to unselected patients. It may generally be desirable to apply both extensive surgery and aggressive combined therapy, but in individual patients such modalities are mostly incompatible. The therapeutic strategy should be tailored according to the malignant potential.

If the mode of malignancy of an individual carcinoma could be determined preoperatively, it would certainly be helpful in considering treatment strategy. Inokuchi et al. [22], in a cytophotometric DNA analysis of early esophageal carcinoma, found that patients with hyperploid tumors exhibited significantly higher recurrence rates than those with near diploid pattern tumors. Table 3 indicates that specimens of the primary tumor of esophageal cancer with node metastasis were found exclusively in DNA pattern types III and IV (hyperploid), although those with no metastasis were distributed equally in all DNA pattern types. Table 4 shows the survival rates of patients with advanced cancer in comparison to those

Table 3. Relationship between DNA patterns and lymph node metastasis (superficial cancer).

DNA patterns	*I*	*II*	*III*	*IV*
Patients without lymph node metastasis	1	8	10	3
Patients with lymph node metastasis	0	2	4	6

Table 4. Five-year survival rates of esophageal cancer in terms of DNA patterns.

DNA patterns	*II*	*III*	*IV*
Superficial cancer	71.1% (7)[a]	71.1% (14)[b]	11.2% (9)[c]
Advanced cancer	47.9% (20)[d]	13.4% (41)[e]	10.4% (38)[f]

a:c, b:c—$P < 0.05$; d:e, d:f—$P < 0.05$; Figures in parentheses indicate number of patients

with superficial cancer, in terms of DNA patterns. In both groups, the difference in survival rate among these types was statistically significant. It should be noted that the survival rate of superficial cancer in type IV was as poor as that in advanced cancer, showing how high the malignant potentiality of type IV is. In contrast, there was a relatively good prognosis in the type II group of advanced cases. Matsuura et al. [23], using biopsy specimens, found that the incidence of lymph node metastasis was higher in types III and IV and that there was a statistical difference compared with type II. In the type IV group, almost all the patients died within 2 years after surgery, whereas patients with type II had a good clinical course, the 5-year survival rate being 60%. Nishihira et al. [24] recently described the significant role of DNA analysis in predicting malignant potential. The DNA distribution pattern is probably a valuable parameter for prediction of the prognosis of squamous cell esophageal cancer.

Concerning type-oriented strategy for treatment, for patients with superficial cancer with a type II or III DNA pattern, or for those without node metastasis, ordinary surgical treatment may be acceptable, but for those with a type IV pattern, perhaps the present approach should be reconsidered. Effective radio-chemo-immunotherapy followed by surgery might be superior to the conventional immediate surgery. In advanced cases in patients without apparent node metastasis, or with a type II pattern, attention should be paid to ways of accomplishing extensive surgery, since this possibly improves the survival. However, in patients with positive node metastasis or type IV patterns, aggressive postoperative combined therapy appears to hold out some hope.

The above statement is only tentative. The control of this presently incurable disease must be aimed at using the most effective method of therapy according to the biological characteristics of the lesion. This may be the key difference between the treatment of esophageal cancer and that of gastric or colon cancer.

References

1. Torek F (1913) The first successful case of resection of the thoracic portion of the esophagus for carcinoma. Surg Gynecol Obstet 16: 614
2. Ohsawa T (1933) Esophageal surgery (in Japanese). J Jpn Surg Soc 34: 1318–1590
3. Adams WE, Phemister DB (1938) Carcinoma of the lower thoracic esophagus: Report of a successful resection and esophago-gastrostomy. J Thorac Surg 7: 621–632
4. Garlock JH (1944) Re-establishment of esophagogastric continuity following resection of esophagus for carcinoma of middle third. Surg Gynecol Obstet 78: 23–28
5. Sweet RH (1945) Transthoracic resection of the esophagus and stomach for carcinoma. Ann Surg 121: 272–284
6. Nakayama K (1950) My own device of operation for thoracic esophageal carcinoma (in Japanese). Nippon Rinsho 8: 289–296
7. Nakayama K (1958) Problems of treatment of thoracic esophageal carcinoma (in Japanese). Nippon Kyobu Geka Gakkai Zashi 6: 637–549
8. Nakayama K, Yanagizawa F, Kato S et al. (1960) Preoperative irradiation of esophageal cancer (in Japanese). Geka (Surgery) 22: 325–328
9. Earlam R, Cunha-Melo JR (1980) Oesophageal squamous carcinoma 1: Critical review of surgery, Br J Surg 67: 381–390
10. Giuli R, Gignoux M (1980) Treatment of carcinoma of the esophagus: Retrospective study of 2400 patients. Ann Surg 192: 44–52
11. Japanese Committee of Registration of Esophageal Cancer: A proposal for a new TNM classification of esophageal carcinoma. (1985) Jpn J Clin Oncol 14: 625–636
12. Japanese Research Society for Esophageal Carcinoma: Report of treatment results of esophageal carcinoma in Japan (1979–1982), No. 10. National Cancer Center, Tokyo, 1990
13. Endo M, Takemoto T, Shirakabe H (1986) Minute lesions of esophageal cancer. Semin Surg Oncol 2: 177–186
14. Kodama Y, Sugimachi K, Soejima K et al. (1981) Evaluation of extensive lymph node dissection of carcinoma of the stomach. World J Surg 5: 241–249
15. Akiyama H, Tsurumaru M, Kawamura T et al. (1981) Principles of surgical treatment for carcinoma of the esophagus: Analysis of lymph node involvement. Ann Surg 194: 438–455
16. Nishihira T, Watanabe T, Ohmori M et al. (1984) Long-term evaluation of patients treated by radical operation for carcinoma of the thoracic esophagus. World J Surg 8: 778–785
17. Kakegawa T (1990) Current progress of treatment of esophageal cancer (in Japanese) Nippon Shokaki Gekagakkai Zasshi (Jpn J Gastroenterol Surg) 23: 685–694
18. Watanabe H (1987) Surgical treatment of thoracic esophageal cancer. In: Proceedings of the 38th Congresso Uruguay Cirujía, Montevideo, 7–11 Dec 1987
19. Orringer MB (1984) Transhiatal esophagectomy without thoracotomy for cancer of the thoracic esophagus. Ann Surg 200: 282–288
20. Shahian DM, Netune WB, and Ellis H (1986) Transthoracic versus extrathoracic esophagectomy: Mortality, morbidity, and long term survival. Ann Thorac Surg 41: 237–246
21. Nishihira T, Hirayama K, Shineha R et al. (1990) Aggressive adjuvant therapy prolongs survival of patients with metastatic carcinoma of the thoracic esophagus. Dis esophagus III: 33–39
22. Inokuchi K, Kuwano H, Sugimachi K et al. (1988) Cytophotometric DNA analysis of surperficial and advanced carcinoma of the esophagus. In: Siewert

JR, Hölscher AH (eds) Diseases of the esophagus—pathophysiology, diagnosis, conservative and surgical treatment. Springer, Berlin Heidelberg, pp 31–34
23. Matsuura H, Sugimachi K, Ueo H et al. (1986) Malignant potentiality of squamous cell carcinoma of the esophagus predictable by DNA analysis. Cancer 57: 1810–1814
24. Nishihira T, Akaishi T, Shineha R et al. (1989) Malignant potential of esophageal carcinoma in terms of biological properties. Dis esophagus II: 125–135

Background of Lymph Node Dissection for Squamous Cell Carcinoma of the Esophagus

Hiroshi Akiyama, Masahiko Tsurumaru, Yoshimasa Ono, Harushi Udagawa, and Yoshiaki Kajiyama[1]

Introduction

Resection of the cancer and reconstruction of a functional organ are the two major components of the surgical treatment of esophageal carcinoma. Although both are equally important, adequate tumor resection is definitely more relevant as far as survival and the operative objectives are concerned, since incomplete resection will make cure impossible. The presence of lymphatic metastases implies the systemic spread of tumor. However, patients with lymph node involvement can still be cured by radical lymph node dissection, as long as the tumor spread remains within certain limits. We have shown [1–4] that radical lymph node dissection in appropriate cases offers a chance of cure in an otherwise uniformly fatal disease.

Diagnostic tools such as ultrasonography and computed tomography generally allow the extent of the tumor to be delineated before operation. However, even with the latest diagnostic techniques, micrometastases in lymph nodes may escape detection. Accordingly, we have proposed the use of routine, systematic lymph node dissection. Candidates should include not only patients with preoperatively detected lymph node metastases, but also those patients whose preoperative evaluation showed no evidence of metastasis. Recent developments in surgical techniques and perioperative care allow an aggressive surgical approach to the management of patients with esophageal carcinoma. Skinner [5] recommended radical en bloc resection because it can be performed with a similar mortality rate to that from ordinary esophagectomy. However, surgery is only a local therapy and resection alone may not always be effective, even in resectable cases. Therefore, it should be kept in mind that resectability and the indications for resection are two different concepts. Obviously, in many of these cases with distant metastases, resection and lymph node dissection are not indicated. On the other hand, even in cases where no metastases are found by the preoperative studies, the potential for distant metastasis is frequently still there. For such cases, the necessity of an interdisciplinary team approach must be emphasized. Current surgical treatment for esophageal cancer cannot be discussed without considering adjuvant

[1] Department of Surgery, Toranomon Hospital, Minato-ku, Tokyo, 105 Japan

therapy. Conversely, adjuvant therapy cannot be evaluated accurately without the perspective provided by a definite surgical strategy.

Lymph Nodes for Removal

Lymphatic Drainage of the Esophagus

The lymphatic drainage of the esophagus is very complicated, with the lymph nodes draining this region being widely distributed from the neck down to the upper abdomen. Pioneering work on the lymphatics of the human esophagus was done in 1903 by Sakata [6], who found that the submucosal lymphatics largely drain in a longitudinal fashion. In the midportion of the esophagus, lymphatics drain to the neck before passing through the muscles to finally reach the nodes, while the lymphatics of the lower thoracic esophagus do not drain to lymph nodes at the same level, but instead drain downwards to the cardiac nodes. The present analysis of lymph node metastases from esophageal cancer still supports the results of this classical study.

Esophageal lymphatics were also studied by Haagensen [7] using sky blue dye injected into the mucosal and submucosal layers of the esophagus during resection of esophageal carcinoma. Observation of the spread of dye within the lymphatics of the resected specimen showed that lymph flowed more freely in the longitudinal direction than in the circumferential direction. This may account for the clinical observation that a considerable length of lymphatic tumor spread is often seen in esophageal cancer, even when patients first present. Haagensen also observed that the inferior esophageal segment mostly empties its lymph into several subdiaphragmatic nodes related to the celiac axis nodes, which lie close to the left gastric and left infradiaphragmatic arteries. Lymph node scanning with ^{99m}Tc sulfur colloid has also been utilized to study esophageal lymphatic drainage [8,9]. Tanabe [10] performed preoperative endoscopic injection of ^{99m}Tc rhenium colloid into the esophageal wall in 42 esophageal cancer patients and then determined the radioactivity of each dissected lymph node with a scintillation counter. He concluded that: (1) the lymphatic flow of the upper and middle thirds of the thoracic esophagus drains mainly into the neck and upper mediastinum and finally from the latter into the abdomen and (2) the lymph from the lower third of the esophagus drains mainly into the abdomen and, in some cases, to the tracheal bifurcation nodes and the nodes around the left renal vein.

Thus, previous studies have shown that the lymphatics of the thoracic esophagus drain mainly into the cervical, mediastinal, and abdominal areas. However, there is usually one predominant area of drainage, depending on the location of the esophageal tumor.

Classification of Lymph Nodes for Surgical Dissection

Anatomical mapping of the lymph nodes of the esophagus was first advocated by Rouvière [11], who classified the different groups of lymph nodes related to the esophagus as follows: (1) the internal jugular vein lymph node chain, (2) the right and left lateral tracheal nodes, (3) the tracheobronchial nodes, (4) the right pulmonary pedicle posterior lymph

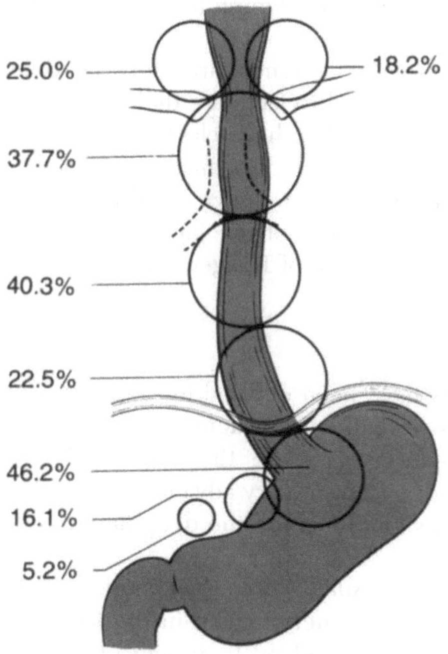

Fig. 1. Gross lymph node regions for dissection for cancer of the thoracic esophagus and frequency of metastases per number of cases. (n = 236)

Table 1. Nomenclature of lymph nodes and frequency of metastasis (per number of cases) in each lymph node group.

		Lymph nodes	*Rates per no. of cases*		
			Right		*Left*
Cervical		Lateral nodes (spinal accessory nerve chain)	2.5%		2.5%
Cervical		Deep external nodes	10.6%		8.9%
Cervical		Deep internal nodes (recurrent nerve chain)	17.8%		10.6%
Mediastinal	Superior	Recurrent nerve chain	19.5%		16.1% (recurrent nerve chain and paratracheal nodes)
Mediastinal	Superior	Paratracheal nodes	5.1%		
Mediastinal	Superior	Brachiocephalic artery nodes		6.8%	
Mediastinal	Superior	Paraesophageal nodes		9.8%	
Mediastinal	Superior	Infra-aortic arch nodes		6.8%	
Mediastinal	Middle	Tracheal bifurcation nodes		15.7%	
Mediastinal	Middle	Pulmonary hilar nodes	7.2%		14.0%
Mediastinal	Middle	Parsesophageal nodes		25.4%	
Mediastinal	Lower	Mediastinal nodes		25.8%	
Mediastinal	Lower	Diaphragmatic nodes (hiatal part)		2.1%	
Abdominal	Superior gastric	Paracardiac nodes		32.6%	
Abdominal	Superior gastric	Lesser curvature nodes		18.2%	
Abdominal	Superior gastric	Left gastric artery nodes		23.7%	
Abdominal		Celiac trunk lymph nodes		16.1%	
Abdominal		Common hepatic artery nodes		5.1%	

n = 236 (Number of cases with complete collo-thoraco-abdominal lymph node dissection, Jan. 1984–Nov. 1991)

nodes, (5) the posterior mediastinal lymph nodes, and (6) the cardiac lymph nodes.

Various classifications and nomenclatures for the different groups of lymph nodes have been devised to facilitate their identification and dissection during resection [12–14]. We have classified these lymph nodes into seven regions, as shown in Fig. 1. The names assigned to the groups of nodes in the different regions are shown in Table 1. This classification is a modification of the guidelines of the Japanese Society for Esophageal Diseases [14] and the description of Haagensen [13].

Cervical Lymph Nodes

Various nomenclatures have been devised for lymph nodes in this region. In patients with cancer of the esophagus, the deep cervical nodes are often involved by metastases, while the superficial cervical nodes are frequently not involved, except in very advanced cases. Therefore, cervical lymph node dissection should include the deep lateral, deep external, and deep internal nodes.

Deep Lateral Nodes (Accessory Nerve Lymphatic Chain). These nodes include those of the accessory nerve chain in the lateral part of the neck. This group is less frequently involved by cancer of the thoracic esophagus and is less important than the deep external and internal nodes.

Deep External Nodes. The nodes external to the internal jugular vein are called the deep external nodes, these include the internal jugular chain and also the nodes of the transverse chain or supraclavicular nodes.

Deep Internal Nodes (Recurrent Laryngeal Nerve Lymphatic Chain). This group includes the nodes medial to the internal jugular vein, the lower part of which is also known as the recurrent laryngeal nerve lymphatic chain. Because of its numerous structural anomalies and variable distribution [15], the right recurrent laryngeal nerve has to be protected carefully while dissecting these lymph nodes. This lymph node group can, in both a theoretical and practical sense, be considered as an en bloc entity with cervical and upper mediastinal portions, and the area where these lymph nodes are located is therefore appropriately named the cervico-thoracic region. This region has a very high incidence of metastasis, so time-consuming and meticulous dissection is necessary. These nodes are the equivalent of the paraesophageal or paratracheal nodes in the neck.

Superior Mediastinal Lymph Nodes

Recurrent Laryngeal Nerve Lymphatic Chain. At the right side of the superior mediastinum, several nodes occur along the right recurrent laryngeal nerve as it winds around the subclavian artery, while on the left side of the superior mediastinum some smaller lymph nodes are located along the left recurrent laryngeal nerve. Since these nodes lie along the left tracheal border, they are also called the left paratracheal nodes. These nodes along the right and left recurrent laryngeal nerves are extremely important where the risk of tumor metastasis and the technical aspects of resection are concerned.

Paratracheal Nodes. The left paratracheal nodes are equivalent to the nodes of the left recurrent laryngeal nerve lymphatic chain, while the right paratracheal nodes are located along the right main trunk of the vagus nerve between the trachea and superior vena cava. The left paratracheal nodes are a frequent site of tumor metastasis, while the right paratracheal nodes are not.

Brachiocephalic Artery Nodes. This group of nodes is located along the brachiocephalic artery, anterior and medial to the vagus nerve where it crosses the artery; their dissection is recommended. These nodes are closely related to the recurrent laryngeal nerve chain and the cervical nodes.

Paraesophageal Lymph Nodes. In the upper mediastinum, the location of the paraesophageal nodes is often controversial. On the left side they are not distinguishable from the paratracheal nodes or the nodes along the left recurrent laryngeal nerve chain. These nodes also include those located in the mediastinum and along the aorta. However, in a strict sense, only those nodes attached to the esophageal wall should be designated as the paraesophageal or periesophageal lymph nodes.

Infra-aortic Arch Nodes. The several nodes located in the infra-aortic arch area may occasionally be the site of metastasis. These nodes are located close to the left recurrent laryngeal nerve when it passes around the aortic arch. These nodes are connected with the tracheobronchial nodes. Some nodes of this group located around Botallo's ligament were described by Engel in 1926 [16] and also by Rouvière in 1932 [11] as Botallo's lymph nodes. These nodes are related to the mid-esophagus and occasionally are the site of metastasis.

Middle Mediastinal Lymph Nodes

Tracheal Bifurcation Nodes. These nodes are also known as the carinal nodes. They are located close to the midesophagus and are a frequent site of metastasis.

Pulmonary Hilar Nodes. The pulmonary hilar nodes are those at the right and left terminal portions of the main bronchi.

Paraesophageal Lymph Nodes. The posterior mediastinal nodes, paraaortic nodes, and nodes along the thoracic duct are included in this category. When these nodes are involved by cancer and infiltration of the surrounding tissues or organs, such as the esophagus, aorta, or thoracic duct, their presence and their relationship with each other become clearer. In the disease-free state, however, these nodes are freely scattered in the sparse mediastinal fat and their specific locations cannot be determined.

Lower Mediastinal Lymph Nodes

Paraesophageal Nodes. The description for the middle mediastinal lymph nodes also applies here.

Diaphragmatic Nodes. These nodes, located at the diaphragmatic level of the lower esophagus, both anterior to the esophagus itself and at the phrenoesophageal angle, form part of the retrocardiac and infracardiac nodes.

Superior Gastric Lymph Nodes

Paracardiac Nodes. These are important nodes, which are frequently involved by esophageal cancer. They are located along the highest branch of the left gastric artery and separately on both sides of the cardia, but they are considered to be a single group because they are all located along the same branch of the left gastric artery. Thus, these nodes are identified as the most cephalad nodes of the gastric lesser curvature.

Lesser Curvature Nodes. The lesser curvature of the stomach is an important route in allowing tumor to spread from the esophagus to the celiac trunk. The majority of the nodes to be removed are along the upper half of the lesser curvature in the left gastric area.

Left Gastric Artery Nodes. Nodes along the left gastric artery are interposed between the root of the artery and the gastric lesser curvature. The paracardiac, lesser curvature, and left gastric artery nodes are important because they are the main routes of tumor spread from the primary lesion to the paraaortic nodes. These can all be considered as nodes of the left gastric area. Nardi and Zuidema [17] reported that lymphatic extension is frequently found in the left gastric nodes and suggested that they should be removed in continuity with the primary lesion.

Celiac Trunk Lymph Nodes. These are the nodes around the short celiac trunk and along the root of the left gastric, common hepatic, and splenic arteries. If lymph node metastasis occurs beyond this point, it indicates systemic dissemination of the tumor.

Common Hepatic Artery Lymph Nodes. These nodes lie along the course of the common hepatic artery and metastases beyond this area also indicate distant tumor spread.

In squamous cell carcinoma of the esophagus, it is rare to find lymph node metastasis in the hepatoduodenal ligament, greater curvature of the stomach, distal splenic artery, or splenic hilum. Therefore, these groups of nodes are not discussed here.

Frequency of Lymph Node Metastasis to Various Sites

Figure 2 shows the lymph node groups that are removed and the extent of resection for cancer of the thoracic esophagus.

The frequency of lymph node metastasis has been investigated according to both tumor location and to the regional location of the lymph nodes themselves. Of 236 patients who underwent collo-thoraco-abdominal dissection, 73.7% had metastatic lymph nodes, including 33.1% with cervical node metastases, 58.1% with mediastinal node metastases, and 49.2% with abdominal node metastases.

Among patients with carcinoma of the upper esophagus, lymph node involvement was recognized in the neck in 44.1%, in the upper mediastinum in 50.0%, in the middle mediastinum in 20.6%, in the lower mediastinum in 5.9%, and in the upper gastric area in 14.7%. Carcinoma of the middle esophagus was associated with lymph node metastasis in 32.9% in the neck, 38.1% in the upper mediastinum, 41.0% in the middle mediastinum, 20.2% in the lower mediastinum, and 42.5% in the upper gastric area. In cases of carcinoma of the lower esophagus, 29.4% of patients had positive lymph node metastasis in the neck, 30.9% had nodes involved in the upper mediastinum, 48.5% in the middle mediastinum, 35.3% in the lower mediastinum, and 69.1% in the upper gastric area. Accordingly, the incidence of metastasis to the neck was high, even when tumors were located in the lower esophagus (Fig. 3).

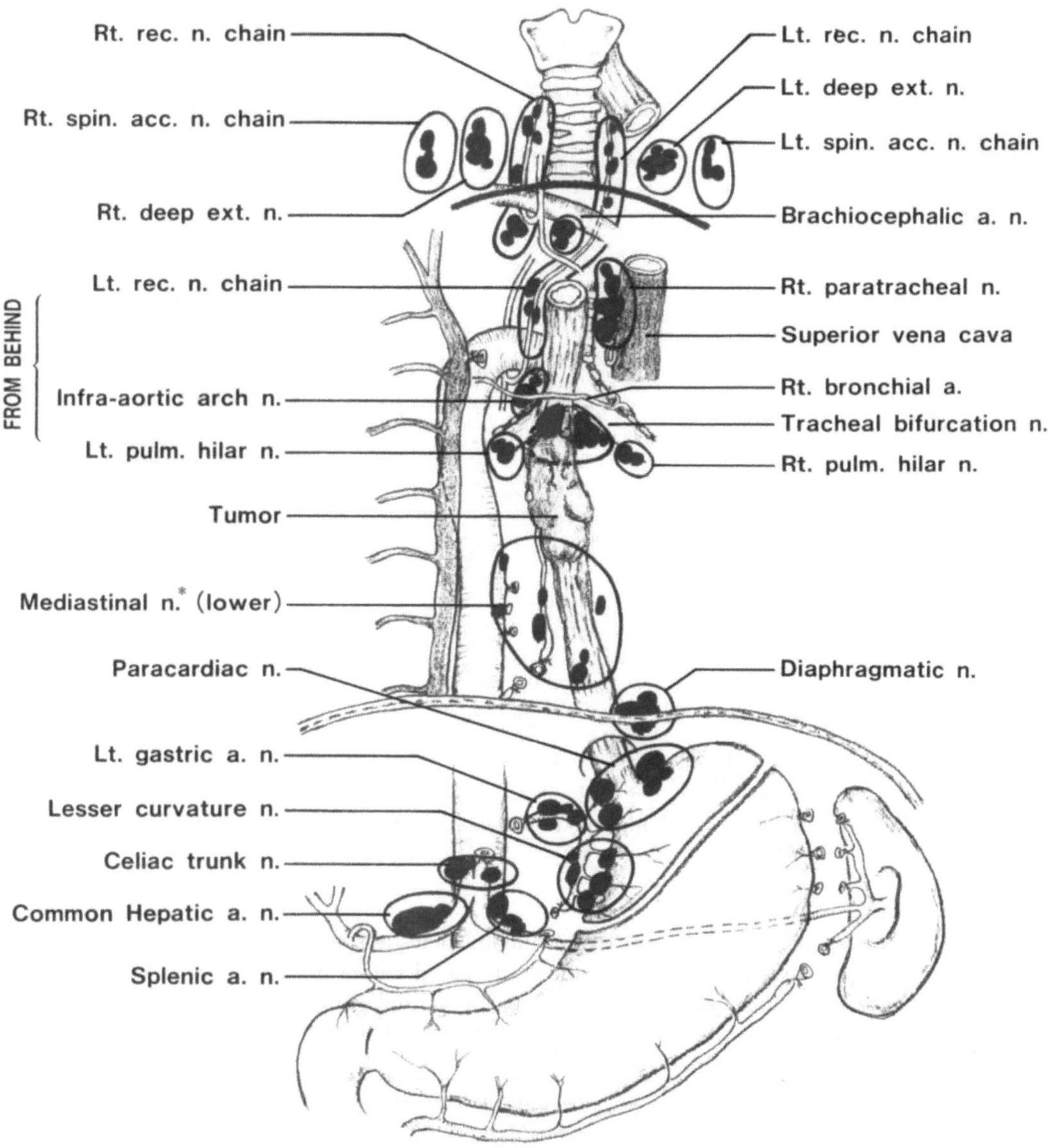

Fig. 2. Extent of resection and collo-thoraco-abdominal lymph node dissection. * Here, mediastinal nodes include paraaortic, thoracic duct, and paraesophageal nodes

Re-Evaluation of Lymph Node Dissection

When patients are properly selected, surgery can provide a definitive means of treating esophageal carcinoma. Radical surgery is limited to local resection and lymphatic clearance, but even the most extensive lymph node dissection will fail once the tumor has spread too far. Thus, to avoid unnecessary or too extensive surgery, the surgeon must have a good understanding of the principles of resection.

Lymph node dissection in surgery for esophageal cancer aims at removing all positive or possibly positive tumor foci in the lymphatic system. The surgical procedure of lymph node dissection may often be the crucial step in saving a patient, since positive nodes are often found among nodes

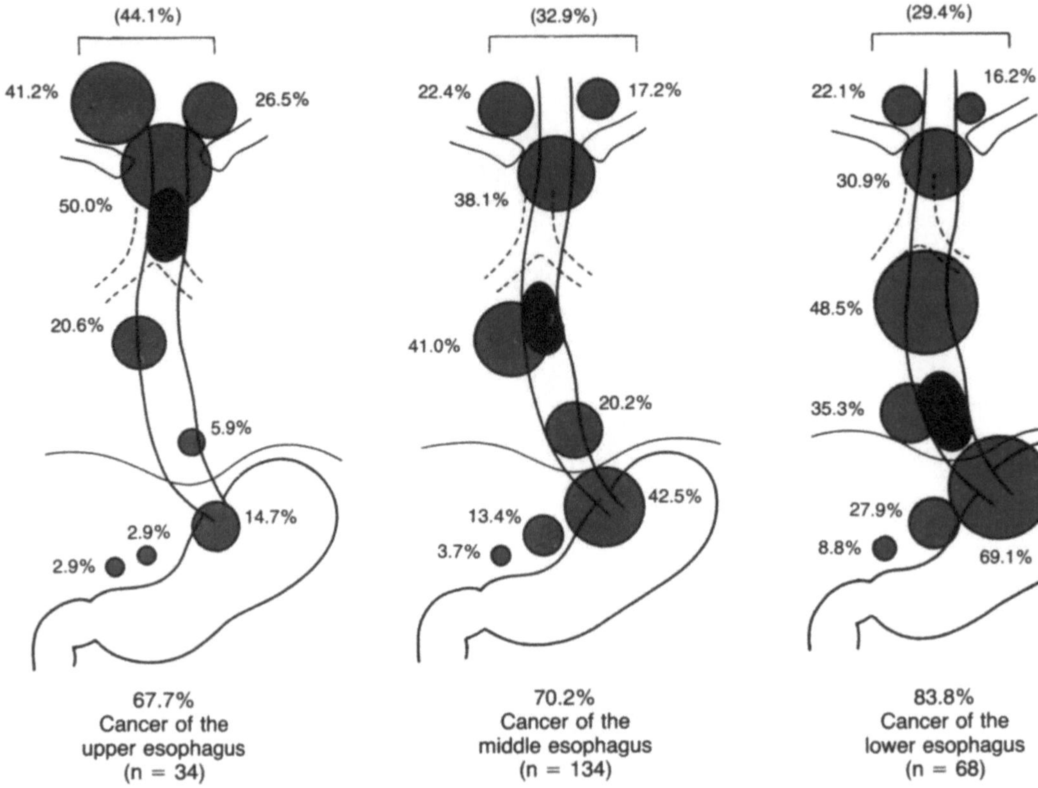

Fig. 3. Frequency of lymph node metastases according to tumor location

Table 2. Frequency of lymph node metastasis in relation to survival time of patients deceased within 5 years and those alive for more than 5 years following initial surgery. (Patients with collo-thoraco-abdominal or thoraco-abdominal lymph node dissection).

Survival time	*Frequency of lymph node metastasis per no. of cases*	
Within 1 year	98/111	(88.3%)
1–2 Years	67/89	(75.3%)
2–3 Years	29/38	(76.3%)
3–4 Years	20/32	(62.5%)
4–5 Years	4/12	(33.3%)
More than 5 years	62/137	(45.3%)

removed from patients who eventually show long-term survival (Table 2). However, surgeons may be discouraged by the rapid recurrence of tumor even after complete removal and so-called "radical" lymph node dissection.

It is extremely difficult to detect all the positive nodes and it is particularly hard to identify micrometastases. Accordingly, the systematic and complete dissection of cervical, mediastinal, and abdominal nodes (collo-thoraco-abdominal dissection) is employed at our institution. Radical lymph

Table 3. Mediastinal and abdominal lymph node metastases and corresponding 5-year survival rates. (Patients without macroscopic residual tumor).

Metastases		*5-Year survival rates*		
Mediastinal nodes	*Abdominal nodes*	*Thoraco-abdominal dissection*	*Collo-thoraco-abdominal dissection**	*p (log rank)*
−	−	52.1%	75.2%	0.0357
−	+	38.7%	60.2%	0.1424
+	−	25.9%	42.7%	0.0821
+	+	13.7%	27.4%	0.0031

* regardless of cervical lymph node metastasis

node dissection, by systematically removing all the nodes which may have macroscopic or microscopic metastases, offers the best hope of a cure. It can never be ascertained whether this procedure is truly preventive, but it must be carried out meticulously because the sites of micrometastasis are always uncertain. There may be some occasions when this procedure is unnecessary, and limited surgery, i.e., lymph node dissection of a lesser extent, can be considered in some selected cases.

It is important for us to continuously re-evaluate the true efficacy of radical lymph node dissection and to establish the appropriate indications and proper extent of this procedure. There are several points about radical lymph node dissection that still cause controversy.

Controversies Concerning Radical Lymph Node Dissection

Anatomical Aspects

Radicality in cancer surgery varies, depending on the organs or tissues and anatomical areas involved. If the lymph node dissection procedures were truly complete and radical, there would be no tumor recurrence in the lymphatic system and all recurrences would be either local or due to hematogenous metastases. In esophageal carcinoma, the lymph nodes that are potential sites of metastasis can be most completely removed from the abdominal area, and dissection tends to be less complete in the cervical and mediastinal regions. In particular, in the cervico-thoracic and superior mediastinal regions, vital organs are intricately related to the lymph nodes, and complete removal of the nodes without traumatizing or resecting these organs is difficult.

Survival rates of groups with and without lymph node metastasis in the mediastinum and abdomen were compared. The 5-year survival rate of patients who had 2-field dissection in the group with negative mediastinal nodes but positive abdominal nodes was 38.7%, but this survival rate dropped to 25.9% in those with positive mediastinal nodes but negative abdominal nodes (Table 3). Thus, involvement of the mediastinal nodes is an extremely important prognostic factor. This may be explained by the fact that mediastinal dissection is anatomically more difficult than epigastric dissection. In the mediastinal area, the upper mediastinum is the most important site, particularly in connection with lymphadenectomy in the cervicothoracic region. As a result of extensive lymphadenectomy in this region, 5-year survival rates have improved significantly (Table 3).

To obtain better operative exposure of the cervico-thoracic region, various surgical approaches have been recommended, but none of them is clearly superior to any other. The authors use a combination of right thoracotomy, upper median laparotomy, and a U-shaped cervical incision as a routine approach.

Does Lymph Node Dissection Cause Tumor Metastasis?

During lymph node dissection, a number of large and small lymph nodes or vessels are traumatized or severed, and this may cause the intra-operative spread of tumor cells. Therefore, lymphatics should be tied, and electrocautery is preferred when lymph nodes and small lymphatic vessels are divided. Resection of certain lymph node groups or the ligation of lymphatics may change the natural lymphatic flow and cause the implantation of tumor cells in unexpected organs or areas. An enthusiastic but careless surgeon may compress the primary lesion or metastatic nodes, resulting in tumor dissemination during the operation. The gentle handling of tissues is thus very important. These problems with lymph node dissection result from mechanical intervention and are unavoidable disadvantages, despite the countermeasures that a competent surgeon will take.

Influence on Morbidity and Mortality

Complete dissection of the three areas of lymph nodes at one time is quite a burden for a patient. On the other hand, esophagectomy without thoracotomy is a simpler operation and causes much less patient morbidity than transthoracic esophagectomy with lymph node dissection. The collo-thoraco-abdominal procedure requires meticulous surgical manipulation of delicate organs and tissues, especially those in the cervico-thoracic region and the superior mediastinum. Also, excessively extensive dissection along the trachea, carina, and bronchi can lead to various postoperative complications. Postoperative hoarseness caused by recurrent laryngeal nerve trauma may be prevented by meticulous dissection and direct visual identification of the full length of the nerve. The extent of the lymph node dissection described is thought to represent the limit of surgical magnitude for poor risk and elderly patients with esophageal cancer. Complete lymph node dissection should always be planned with an appropriate balance between the curative value of the operation and the risks it poses to the patient.

Removal of a Cancer Barrier

Removal of involved lymph nodes is accepted as necessary for the cure of the disease. However, the removal of normal nodes is obviously a disadvantage because it only destroys a barrier to the spread of cancer.

The Immunosuppressive Effect of Surgery

Since complete lymph node dissection implies extensive surgery, it can have an immunosuppressive effect, and a small residual hidden nest of tumor cells may become manifest after major surgery for this reason.

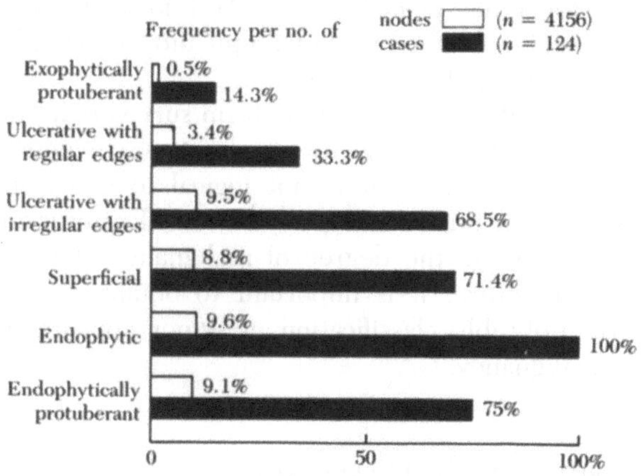

Fig. 4. Frequency of lymph node metastases according to macroscopic pathology of primary tumors

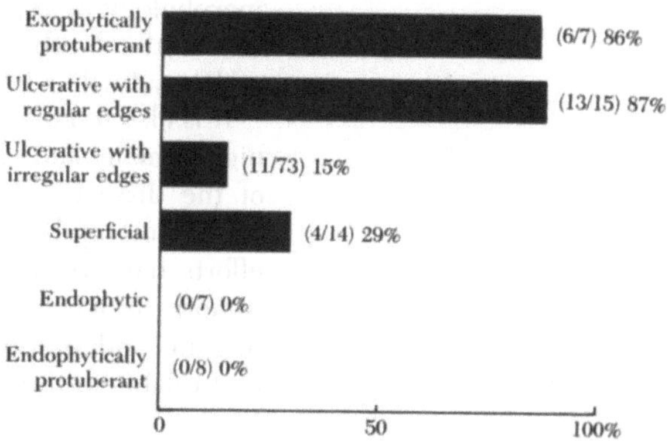

Fig. 5. Five-year survival rates for different types of esophageal tumors after resection (124 cases with minimum 5-year follow up)

Can a Patient Survive with Positive Nodes?

Logically, long-term survival is not considered possible for a patient with unresected positive lymph nodes. However, in some circumstances tumor spread can be controlled for a considerable period with the aid of adjuvant therapy. Comparison of survival between two groups with and without lymph node dissection is usually impossible because of differences in the nature of both tumors and hosts.

Is Radical Surgery Contraindicated for Highly Malignant Tumors?

Regardless of the size and depth of the cancer, tumors with a highly malignant nature often have widespread lymphatic or hematogenous metastases at the time of detection of the disease or at a later date. When metastases are found preoperatively in all three of the cervical, mediastinal, and abdominal regions, long survival cannot be expected, even with complete lymph node dissection.

Relationship Between Tumor Morphology and Malignancy

Morphological growth patterns were correlated with the survival rate after resection in 124 patients with squamous cell carcinoma. The 124 patients were divided into six groups according to the pattern of tumor growth observed at endoscopy and subsequently verified by examination of the resected specimens. These patterns were: (1) exophytically protuberant, (2) ulcerative with regular edges, (3) ulcerative with irregular edges, (4) superficial, (5) endophytic, and (6) endophytically protuberant. There were close correlations between the macroscopic appearance of the primary tumor and the frequency of lymph node metastasis (Fig. 4). Regardless of the extent of lymphatic spread at the time of surgery, the 5-year survival rates after resection for each tumor pattern were: (1) 86%, (2) 87%, (3) 15%, (4) 29%, (5) 0%, and (6) 0%, respectively (Fig. 5). Exophytically protuberant lesions and ulcerative tumors with regular edges were the least malignant, while ulcerative tumors with irregular edges and superficial tumors were moderately malignant. The endophytic and endophytically protuberant types were highly malignant. It has been proposed that the preoperative staging of esophageal cancer should be based on tumor morphology as representing the grade of malignancy. Additional information on the depth of invasion can be obtained by endoscopic ultrasonography and may further improve the staging of this disease.

It is also important to realize that there are no long-term survivors in the highly malignant group, even when microscopy shows no involvement of the dissected lymph nodes. Therefore, it is not logical to carry out complete lymph node dissection for highly malignant tumors. Enthusiastic efforts have been made to determine the degree of malignancy of these tumors (Inokuchi [18]). In the future, it is important to obtain a more definite and preoperatively applicable classification of tumor appearance that relates to the grade of malignancy.

Strategy for Radical Lymph Node Dissection

Huang [19] has stated that major improvement in survival has been achieved by detection of this disease in its early stages, and Ellis et al. [20] stated that "It is concluded that long-term survival of patients with carcinoma of the esophagus or cardia will probably not improve until early diagnosis is possible."

The importance of early diagnosis cannot be overemphasized. If the tumor is found as a carcinoma in situ, treatment is obviously much simpler. However, the majority of patients are still diagnosed in the advanced stage. In advanced but operable cases, en bloc resection or complete lymph node dissection have obtained relatively favorable survival rates. The Japanese National Survey showed that the operative method generally included removal of the peri-esophageal and para-esophageal lymph nodes. The overall 5-year survival rate in this series was 23.1% [21], while that of the author's (Akiyama) series (radical lymph node dissection) was 34.6% [4]. The difference between the two series suggests that there is a favorable effect of radical lymph node dissection in patients both with and without macroscopic node involvement.

The authors compared the survival rates for 523 patients without macroscopic residual tumor who were treated by collo-thoraco-abdominal dissection or by thoraco-abdominal dissection. The 5-year survival rate after collo-thoraco-abdominal dissection was 48.0%, while that after the thoraco-abdominal approach was 35.6% (Fig. 6a). Some subgroups were analyzed to better evaluate the results. A significant difference was seen between the subgroups of patients with positive mediastinal and abdominal nodes M[+]A[+] (Fig. 6b) and in the subgroup of patients with negative mediastinal and abdominal nodes (M[−]A[−]) (Fig. 6c). The operative mortality rate within 30 days after 523 thoraco-abdominal or collo-thoraco-abdominal dissections was 2.5% and the postoperative hospital death rate (including death within 30 days and death due to early recurrence) was 5.2%. However, in 84 consecutive cases over the last two years, the rates have been improved to 1.2% and 2.4%, respectively.

The most typical operation without lymph node dissection for esophageal cancer is esophagectomy without thoracotomy. Orringer [22] reported 100 cases treated by this method; the operative mortality was 6% and the 5-year survival rate was 17%. He stated that these figures are as good as those reported in many series of transthoracic esophagectomies. However, a comparison of these results with ours suggests that complete lymph node dissection plays an effective role in increasing survival. Nevertheless, it is

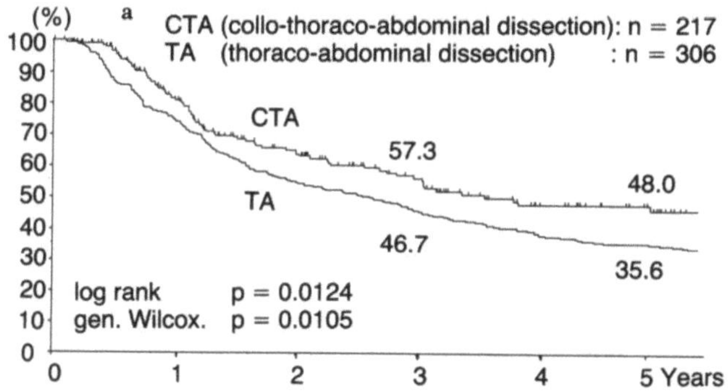

Fig. 6a–c. Comparison between the survival rates of groups with collo-thoraco-abdominal dissection (CTA) and thoraco-abdominal dissection (TA). **a** Overall survival rates. **b** Survival rates for patients with positive mediastinal and abdominal nodes M[+]A[+]. **c** Survival rates for patients with negative mediastinal and abdominal nodes M[−]A[−]

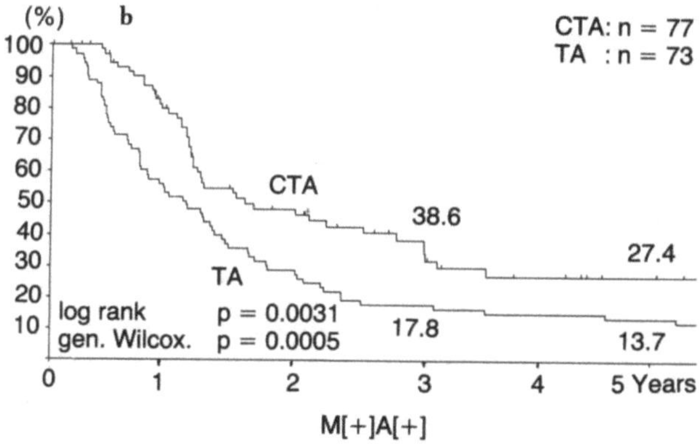

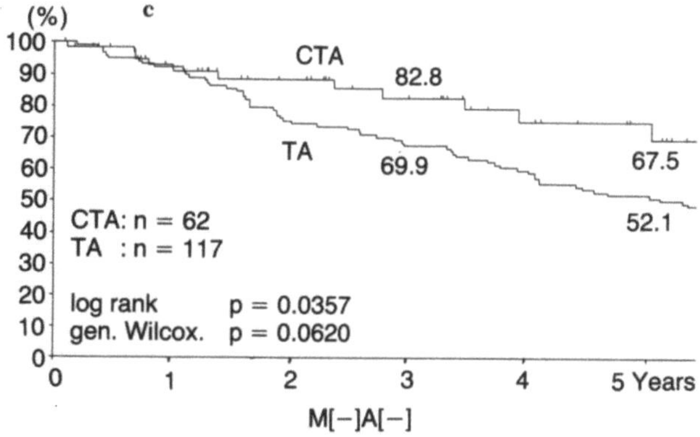

extremely difficult to compare the results of different reports by accurate statistical analysis. Even the conventional staging data for the patients is often not available in many reports. When esophagectomy without thoracotomy is performed, we have no data concerning tumor spread. Furthermore, esophagectomy without thoracotomy is not performed for locally advanced disease. Although radical lymph node dissection seems to be a useful means of improving the results in the majority of cases, it is not applicable to all patients; our strategy is outlined below.

Cases Where Dissection Is Unnecessary

It is known that cancer in situ is confined to the epithelium and has no lymphatic spread. Therefore, lymph node dissection is not necessary in such cases, in which esophagectomy without thoracotomy or endoscopic stripping of the esophagus is indicated. When the cancer in situ is a single, small lesion, non-surgical methods such as laser therapy may also be chosen.

As described previously, there are particular tumor patterns associated with low malignancy, e.g., exophytically protuberant tumors with a cauliflower appearance. When this type of tumor is small and at a comparatively early stage, lymph node dissection is thought to be unnecessary. However, at present this cannot be said with certainty and it is still up to the individual judgement of the surgeon.

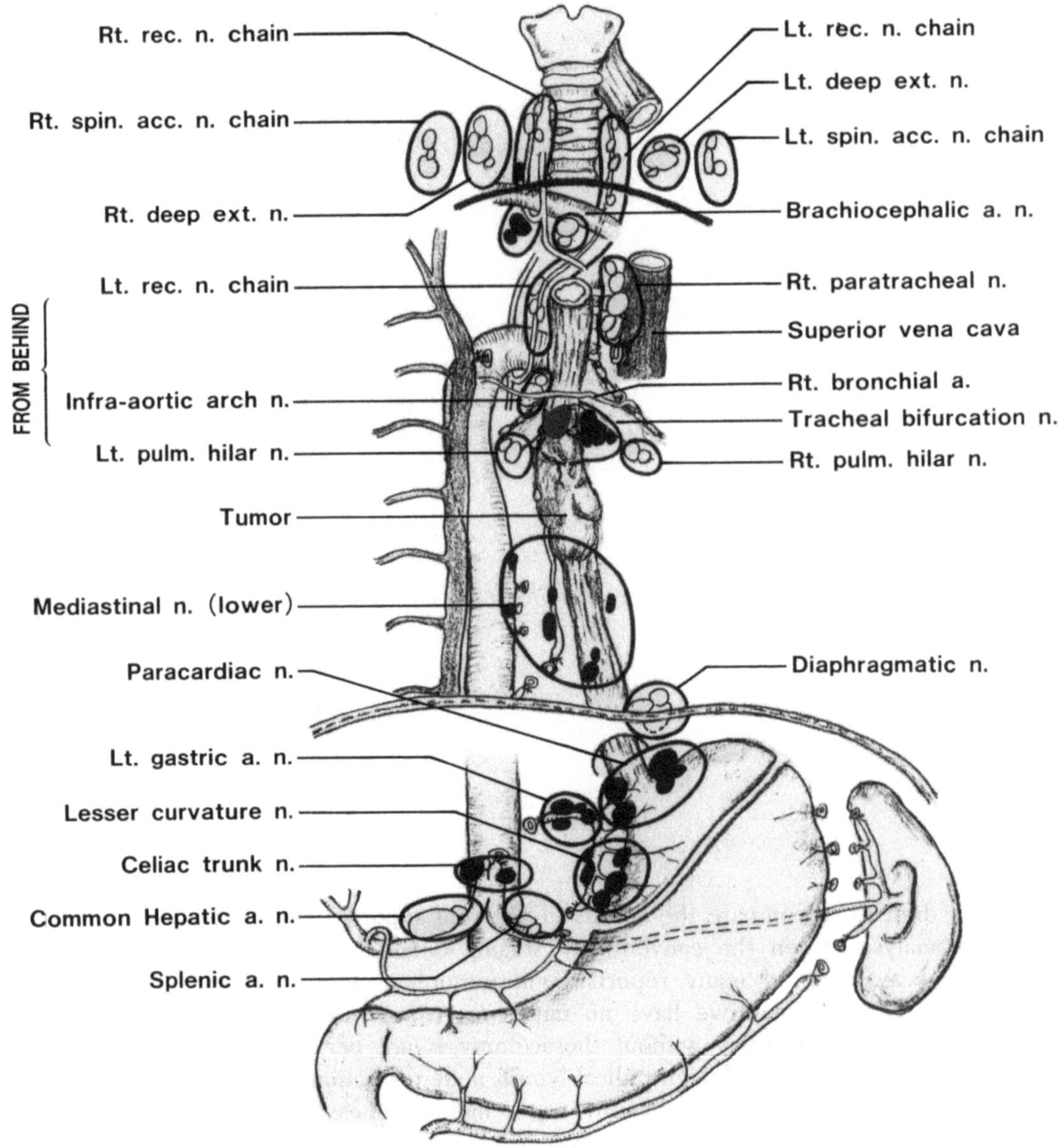

Fig. 7. Extent of resection and less extensive lymph node dissection. Nodes marked in black are dissected

Indications for Dissection

Once the tumor penetrates the muscularis mucosae, lymphatic spread may occur and lymph node dissection is necessary. For cancers with low malignancy (except for the earliest stage) and for cancers with a moderate degree of malignancy, complete lymph node dissection is indicated, and this significantly improves the long-term results.

Contraindications to Dissection

For highly malignant cancers, lymph node dissection is useless and is not indicated. Examples of such tumors are those with an endophytically protuberant pattern or highly malignant cancers manifested by multiple intramural satellite nodules. These tumors have a very poor prognosis, regardless of the status of the lymph nodes at the time of presentation. In this group, even when the dissected nodes are all negative for cancer, the outcome is not improved. Similarly, in the case of malignant melanoma, lymph node dissection not only does not produce any significant improvement, but can even be deleterious [23,24]. Needless to say, node dissection is not indicated in the case of local palliative resection or for patients with distant metastases.

In general, radical lymph node dissection is useful to improve survival and should be applied according to the indications described above. For selected patients, such as those with early stage disease or those who are poor surgical risks, less extensive node dissection, an example of which is shown in Fig. 7, is recommended.

References

1. Akiyama H (1980) Surgery for carcinoma of the esophagus. In: Current problems in surgery. Year Book Medical, Chicago, p 69
2. Akiyama H, Tsurumaru M, Kawamura T, Ono Y (1981) Principles of surgical treatment for carcinoma of the esophagus. Ann Surg 194: 438–446
3. Akiyama H (1988) Cardinals in the regional lymph node dissection in surgery of thoracic esophageal cancer. In: Siewert JR, Hölscher AH (eds) Diseases of the esophagus. Springer, Berlin, p 416
4. Akiyama H (1990) Surgery for cancer of the esophagus. Williams and Wilkins, Baltimore
5. Skinner DB (1983) En bloc resection for neoplasms of the esophagus and cardia. J Thorac Cardiovasc Surg 85: 59–71
6. Sakata K (1903) Über die Lymphgefässe des Oesophagus und über seine regionalen Lymphdrüsen mit Berücksichtigung der Verbreitung des Karzinoms. Mitt Grenzgeb Med Chir 11: 634–656
7. Haagensen CD (1972) The lymphatics in cancer. WB Saunders, Philadelphia
8. Terui S, Kato H, Iizuka T (1982) An evaluation of the mediastinal lymphoscintigram for carcinoma of the esophagus studied with ^{99m}Tc Rhenium sulfur colloid. Eur J Nucl Med 7: 99–101
9. Hauser W, Atkins HL, Richards P (1969) Lymph node scanning with ^{99m}Tc sulfur colloid. Radiology 92: 1369–1371
10. Tanabe G, Baba M, Kuroshima K, Natsugoe S, Yoshida H, Aikou T, Kajisa T (1986) Clinical evaluation of esophageal lymph flow system based on the RI uptake of removed regional lymph nodes following lymphoscintigraphy (in Japanese). J Jpn Surg Soc 87: 315–323
11. Rouvière H (1932) Anatomie des lymphatiques de l'homme. Masson et Cie, Paris
12. Sweet RH (1950) Thoracic surgery. WB Saunders, Philadelphia, pp 228–229
13. Haagensen CD (1972) The lymphatics in cancer. WB Saunders, Philadelphia, pp 60–84
14. Japanese Society for Esophageal Diseases (1976) Guidelines for clinical and pathologic studies on carcinoma of the esophagus. Part I. Clinical classification. Jpn J Surg 6: 69–78
15. Gray SW, Skandalakis JE (1985) Atlas of surgical anatomy for general surgeons. Williams and Wilkins, Baltimore, pp 37–38
16. Engel S (1926) Die Topographie der bronchialen Lymphknoten und ihre präparatorische Darstellung. Beitr Klin Tuberkulose 64: 468–481
17. Nardi GL, Zuidema GD (1965) Surgical treatment of advanced esophageal carcinoma. Am J Surg 109: 583–586

18. Inokuchi K (1988) Milestones in the treatment of esophageal cancer along the road to improved results. Dis Esophagus 1: 13–18
19. Huang GJ (1981) Early detection and surgical treatment of esophageal cancer. J Jpn Surg Soc 11: 399–405
20. Ellis FH, Gibb SP, Watkins E (1983) Esophagogastrectomy. A safe, widely applicable and expeditious form of palliation for patients with carcinoma of the esophagus and cardia. Ann Surg 198: 531–540
21. Japanese Committee of Registration of Esophageal Cancer (1985) A proposal for a new TNM classification of esophageal carcinoma. Jpn J Clin Oncol 14: 625–636
22. Orringer MB (1984) Transhiatal esophagectomy without thoracotomy for cancer of the thoracic esophagus. Ann Surg 200: 282–288
23. Veronesi U (1987) Delayed node dissection in stage one malignant melanoma: Justification and advantages. Cancer Invest 5: 47–53
24. Abdi EA, Hanson J, McPherson TA (1987) Adjuvant chemoimmunotherapy after regional lymphadenectomy for malignant melanoma. Am J Clin Oncol 10: 117–122

Illustrations and Photographs of Surgical Esophageal Anatomy Specially Prepared for Lymph Node Dissection

TATSUO SATO and KAZUHIRO SAKAMOTO[1]

Introduction

In radical operations for cancer, removal of the related lymph nodes, in addition to the removal of the organ, is considered to be essential. Detailed and comprehensive anatomical knowledge of the regional lymph nodes, their relationship to the surrounding structures, and the routes of lymph flow from these nodes to the venous angles is of primary importance. Thus, it is said that the surgery of cancer is not only the surgery of organs, but also the anatomy of the lymphatic system.

The lymphatics of the mediastinum and neck, which are associated with the esophagus and run along the trachea, were dissected in detail in several anatomical specimens. In order to highlight the lymph nodes in this area, several illustrations and diagrams, based on the actual dissection, were drawn. The actual photographs taken during the dissection are also included for a more complete understanding of the intricate topographical relationships.

Although there are numerous transverse connections, the mediastinal lymphatics can be roughly divided into two major pathways: the right pathway and the left pathway. Each pathway may then be subdivided into a superficial pathway along the major blood vessels and a deep pathway along the trachea and esophagus. Here, in order to more clearly understand the location and communication of the complicated mediastinal lymph nodes, these pathways are used to facilitate the detailed description.

Regarding the terminology of the mediastinal lymph nodes, although numerous systems of organized nomenclature have been proposed, there has not been an agreement to date on any one set of terms. Table 1 is included as a reference for the terminology used in this paper and for comparison with the terminology used in well-known reports in the literature [1–13].

[1] Second Department of Anatomy, Faculty of Medicine, Tokyo Medical and Dental University, Tokyo, 113 Japan

Table 1. Comparative terminology of lymph nodes related to esophageal cancer (lower cervical and upper and middle thoracic nodes).

Poirier and Cunéo (1902) [1]	*Bartels (1909) [2]*	*Rouvière (1932) [3]*	*Rouvière and Tobias (1938) [4]*	*Kutsuna (1968) [5]*	*Haagensen et al. (1972) [6]*
Chaîne cervicale profonde	Lgll. cervicales profundi inferiores	Chaîne jugulaire interne	Internal jugular chain	Lnn. cervicales profundi	Internal jugular chain
Ganglions sus-claviculaires	Lgll. supraclaviculares			Lnn. supra-claviculares	
Chaîne antérieure profonde	Lgll. cervicales anteriores profundae	Ganglions cervicaux antérieurs (ganglions juxta-viscéraux)	Anterior cervical nodes (juxta-visceral nodes)	Lnn. cervicales anteriores profundi	Anterior cervical nodes (juxtavisceral chain)
Amas prétracheal	Lgll. praetracheales	Group prétrachéal	Pretracheal group	Lnn. pretracheales	Pretracheal nodes
Chaîne récurrentielle	Lgll. paratracheales (Recurrenskette)	Group latérotrachéal (chaîne récurrentielle)	Laterotracheal group (recurrent chain)	Lnn. paratracheales	Recurrent nerve chain
					Highest anterior mediastinal nodes
Ganglions médiastinaux antérieurs	Lgll. mediastinales anteriores	Ganglions médiastinaux antérieurs (ou prévasculaires)	Anterior mediastinal nodes	Lnn. mediastinales anteriores (superiores)	Anterior mediastinal nodes
	Lgll. anguli anonymi	Chaîne ganglionnaire médiastinale antérieure droite	Right anterior mediastinal chain	Lnn. anguli brachiocephalici	
		Chaîne ganglionnaire médiastinale antérieure transversale	Transverse anterior mediastinal chain		
		Chaîne ganglionnaire médiastinale antérieure gauche	Left anterior mediastinal chain	Lnn. arcus aortae	
		Ganglion canal de Botal	Node of the duct of Botallo		
Ganglions péritrachéo-bronchiques	Lgll. bronchiales	Ganglions péritrachéaux	Peri-tracheal nodes	Lnn. tracheales	Paratracheobronchial nodes (middle mediastinal nodes)
Ganglions trachéo-bronchiques droits	Lgll. paratracheales dextrae	Chaîne latérotrachéale (ou paratrachéale) droite	Right laterotracheal (or paratracheal) chain	Lnn. paratracheales dextri	Right paratracheal nodes
	Lgll. tracheobronchiales dextrae			Lnn. tracheo-bronchiales superiores dextri	Right lateral paratracheobronchial nodes
Ganglions trachéo-bronchiques gauches	Lgll. paratracheales sinistrae	Chaîne latérotrachéale (ou paratrachéale) gauche	Left laterotracheal (or paratracheal) chain	Lnn. paratracheales sinistri	Left paratracheal nodes
	Lgll. tracheobronchiales sinistrae			Lnn. tracheo-bronchiales superiores sinistri	Left lateral paratracheobronchial nodes
Ganglions intertrachéo-bronchiques	Lgll. bifurcationis	Ganglions de la bifurcation (ganglions intertrachéo-bronchiques)	Nodes of the tracheal bifurcation (inter-tracheobronhial nodes)	Lnn. tracheo-bronchiales inferiores	Subcorynal nodes
Ganglions interbronchiques	Lgll. broncho-pulmonares	Ganglions des pédicules pulmonaires Ganglions antérieurs Ganglions postérieurs Ganglions supérieurs	Nodes of the pulmonary roots Anterior nodes Posterior nodes Superior nodes	Lnn. broncho-pulmonares	
		Ganglions inférieurs (ganglions du ligament triangulaire)	Inferior nodes (nodes of the ligamentum latum pulmonis)		Nodes of the pulmnonary ligament
Ganglions médiastinaux postérieurs	Lgll. mediastinales posteriores	Ganglions médiatinaux postérieurs	Posterior mediastinal nodes	Lnn. mediastinales posteriores	Posterior mediastinal nodes
		Ganglions juxta-œsophagiens	Juxta-oesophageal nodes		
		Ganglions interaortico-œsophagiens	Interaortico-oesophageal nodes		Para-aorto-esophageal nodes
		Ganglions juxta-aortiques	Juxta-aortic nodes		

Nomina Anatomica, 5th edn (1983) [7]	*Guidelines for Clinical and Pathologic Studies on Carcinoma of the Esophagus, 6th edn (1984) [8]*	*Clemente (1985) [9] (Gray's Anatomy, 30th American edition)*	*Akiyama (1990) [10]*	*Miscellaneous*	*Present paper*
Lnn. cervicales laterales profundi	Deep cervical lymph nodes	Deep cervical nodes	Deep external cervical nodes		Internal jugular nodes
Lnn. supraclaviculares	Supraclavicular lymph nodes				Supraclavicular nodes
Lnn. cervicales anteriores profundi		Anterior cervical nodes			Deep anterior cervical nodes
Lnn. pretracheales		Pretracheal nodes			Pretracheal nodes
Lnn. paratracheales		Paratrached nodes	Recurrent nerve chain		Paratracheal nodes (recurrent nerve chain)
	Cervical paraesophageal lymph nodes				Periesophageal nodes
			Right recurrent nerve chain (lowest)	Right highest upper mediastinal nodes (Kinoshita et al.) [11] TOP nodes (Nishihira et al.) [12]	Right recurrent nerve nodes (highest mediastinal nodes)
Lnn. mediastinales anteriores		Anterior mediastinal nodes			Anterior mediastinal nodes
			Nodes on the brachiocephalic artery		
Nodus ligamentis arteriosi			Nodes around the ligamentum arteriosum (Infra-aortic arch nodes)	Lgl. ductus Botalli (Engel) [13]	Nodes of the ligamentum arteriosum
Lnn. paratracheales	Thoracic paratracheal lymph nodes	Tracheobronchial nodes			
Lnn. paratracheales dextri		Right tracheal nodes	Right paratracheal nodes		Right paratracheal nodes
Lnn. tracheobronchiales superiores dextri (nodus arcus venae azygos)		Right superior tracheobronchial nodes			
Lnn. paratracheales sinistri		Left tracheal nodes	Left paratrachial nodes		Left paratracheal nodes
Lnn. tracheobronchiales superiores sinistri		Left superior tracheobronchial nodes	Left recurrent nerve chain		Infraaortic arch nodes
Lnn. tracheobronchiales inferiores	Bifurcation lymph nodes	Inferior tracheobronchial nodes	Tracheal bifurcation nodes	Lgl. epibronchiales sinistri (Engel) [13]	Infracarinal nodes
	Pulmonary hilar lymph nodes	Bronchopulmonary nodes	Pulmonary hilar nodes		Bronchopulmonary nodes
Lnn. mediastinales posteriores		Posterior mediastinal nodes	Posterior mediastinal nodes		Posterior mediastinal nodes
Lnn. juxta-esophageales	Thoracic paraesophageal lymph nodes (upper, middle and lower)		Para-oesophageal nodes		Periesophageal nodes
	Posterior mediastinal lymph nodes				

Figure Abbreviations

A, aorta
AA, arch of the aorta
B, bronchus
BT, bifurcation of the trachea
CV, cervical vertebra
D, diaphragm
E, esophagus
F, connecting fascia between the prevertebral fascia and the carotid sheath
L, lung
LX, left vagus nerve
P, pleura
PC, pericardium
PH, pharynx
RP, retropharyngeal space
RX, right vagus nerve
SVC, superior vena cava
T, trachea
TH, thyroid gland
VC, vertebral column
X, vagus nerve

az, azygos vein
ba, bronchial artery
bc, brachiocephalic trunk
ca, common carotid artery
ea, esophageal artery
ib, interbronchial membrane
ic, intercostal artery
it, internal thoracic artery
jv, internal jugular vein
la, ligamentum arteriosum
lb, left brachiocephalic vein
lr, left recurrent laryngeal nerve
oh, omohyoid muscle
pa, pulmonary artery
ph, phrenic nerve
pt, pulmonary trunk
pv, pulmonary vein
rb, right brachiocephalic vein
rr, right recurrent laryngeal nerve
sa, subclavian artery
sc, scalenus anterior muscle
sv, subclavian vein
st, sympathetic trunk
ta, inferior thyroid artery
td, thoracic duct
te, tracheoesophageal muscle
vv, vertebral vessels

ban, brachiocephalic angle nodes
ian, infraaortic arch nodes
icn, infracarinal nodes
l, lymph vessel
lan, nodes of the ligamentum arteriosum
ltn, left paratracheal nodes
pen, periesophageal node
ptn, pretracheal part of the paratracheal nodes
rnn, right recurrent nerve nodes
rtn, right paratracheal nodes
scn, supraclavicular nodes

Fig. 1

This is a scheme showing the nodes along the right and left pathways and their close relationships to the trachea, major vessels, and nerves. In order to clearly view the trachea and bronchi, the heart and major vessels have been removed and the superior vena cava has been shifted to the right. The ascending lymphatics in the pulmonary ligaments and along the lower thoracic esophagus and the bronchi converge at the nodes of the bifurcation area. These nodes are divided into three communicating groups: infracarinal nodes (nodes of the tracheal bifurcation), infraaortic arch nodes (left tracheobronchial nodes), and azygos arch nodes (right tracheobronchial nodes). The ascending lymphatics along the trachea and esophagus are roughly divided into right and left pathways. Both pathways have deep and superficial routes, according to their relationship with the major vessels, which drain into the nodes close to their respective venous angle.

Fig. 1. Major pathways of the paratracheal lymph nodes related to the esophagus reaching the right and left venous angles

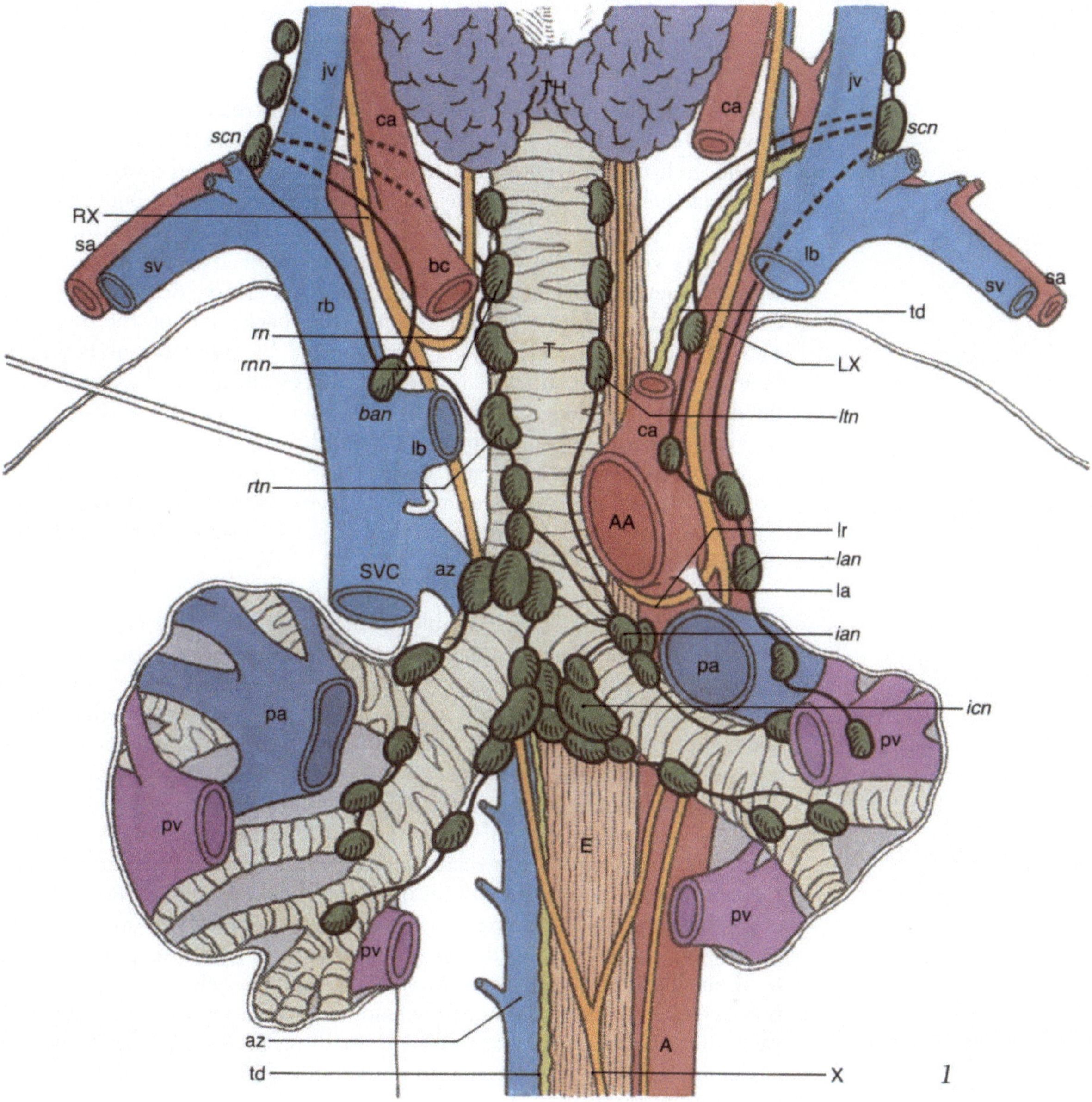
jv
ca
TH
ca
jv
scn
scn
RX
sa
sv
bc
lb
sv
sa
td
rb
rn
rnn
T
LX
ltn
ban
lb
ca
rtn
AA
lr
lan
SVC
az
la
ian
pa
pa
icn
pv
pv
E
pv
pv
az
A
td
X

1

Right Deep Pathway

Fig. 2

This is an illustration of the esophagus and related structures when approached from the right side, based on the dissection of specimen 1. As the arch of the azygos vein has been cut and moved away, the whole length of the thoracic esophagus can be seen. The lung has been shifted to the front. Thus, the back surface of the right bronchus as well as the inferior pulmonary vein are visible. The vagus nerve, after giving off the recurrent laryngeal nerve which surrounds the subclavian artery, descends along the trachea slightly obliquely and reaches the back side of the bronchus. Thus, this quadrilateral tracheal area (formed by the superior vena cava, brachiocephalic and subclavian arteries, and esophagus and azygos vein) is divided into two long triangles by the vagus nerve. From the highest intercostal artery of aortic origin, the right bronchial artery arises and passes over the esophagus and under the vagus nerve, extending to the bronchus and lung. The connective tissues surrounding various nodes have been removed to obtain a clearer view, making each node more distinguishable.

Many lymph nodes are visible; some nodes are located behind the pericardium. Those under the bifurcation of the trachea are the infracarinal nodes. Along the bronchial artery from the intercostal artery, two nodes are observable: one is located at the origin of this artery and the other is above the border between the trachea and esophagus. Several nodes are located along the trachea. Most of these are anterior to the vagus nerve; however, a few are posterior to the nerve and near the esophagus. Two nodes are situated behind and under the subclavian artery near the esophagus, adjacent to the recurrent laryngeal nerve. These are the right recurrent nerve nodes, sometimes designated the highest mediastinal nodes or the lowest nodes of the right recurrent nerve chain. These are thought to occupy an important position in the lymphatics of the mediastinum. Due to their location, these nodes receive lymph vessels from most of the paratracheal nodes, including the chains both anterior and posterior to the vagus nerve. The ascending vessels from these nodes continue to the root of the neck and drain directly into the venous angle or indirectly via nodes close to this angle.

Fig. 2. Lymph nodes along the right margin of the trachea and esophagus in the thorax (specimen 1, male)

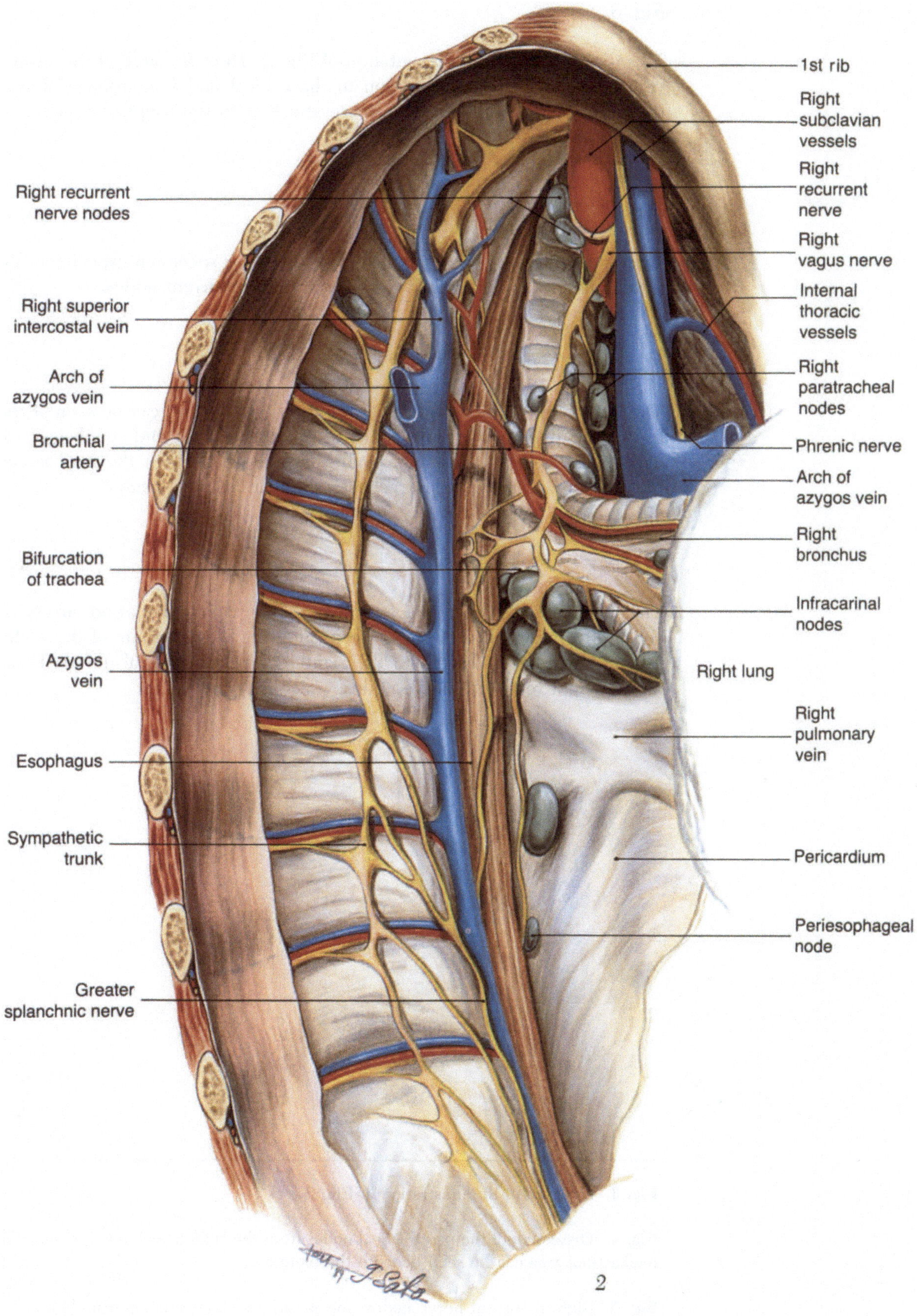
1st rib
Right subclavian vessels
Right recurrent nerve
Right vagus nerve
Internal thoracic vessels
Right paratracheal nodes
Phrenic nerve
Arch of azygos vein
Right bronchus
Infracarinal nodes
Right lung
Right pulmonary vein
Pericardium
Periesophageal node
Right recurrent nerve nodes
Right superior intercostal vein
Arch of azygos vein
Bronchial artery
Bifurcation of trachea
Azygos vein
Esophagus
Sympathetic trunk
Greater splanchnic nerve

2

Fig. 3

This is a diagramatic representation of Fig. 2. Here the arch of the azygos vein is shown intact. In addition to the critical nodes mentioned above, the periesophageal nodes, which are attached to the esophagus, are also shown.

Figs. 4–7

This is a series of photographs of a different specimen (specimen 2), which offer a supplementary view of the right recurrent nodes.

Fig. 4

The right lung has been removed and the thyroid gland can be seen at the *top* of the photograph. The right side of the trachea and esophagus are visible, covered by the mediastinal pleura. Also through the mediastinal pleura, two anthracotic paratracheal nodes can be seen (*center*).

Fig. 5

The mediastinal pleura was removed. The right subclavian artery is reflected upwards by a Pean's forceps allowing view of some of the nodes behind the artery adjacent to the right recurrent laryngeal nerve. These are the right recurrent nerve nodes.

Fig. 3. Diagram of the illustration in Fig. 2

Fig. 4. Dissection of the paratracheal nodes from the right (specimen 2, male). The mediastinal area can be seen beneath the pleura

Fig. 5. Right recurrent nerve nodes (specimen 2) with permission from [14]

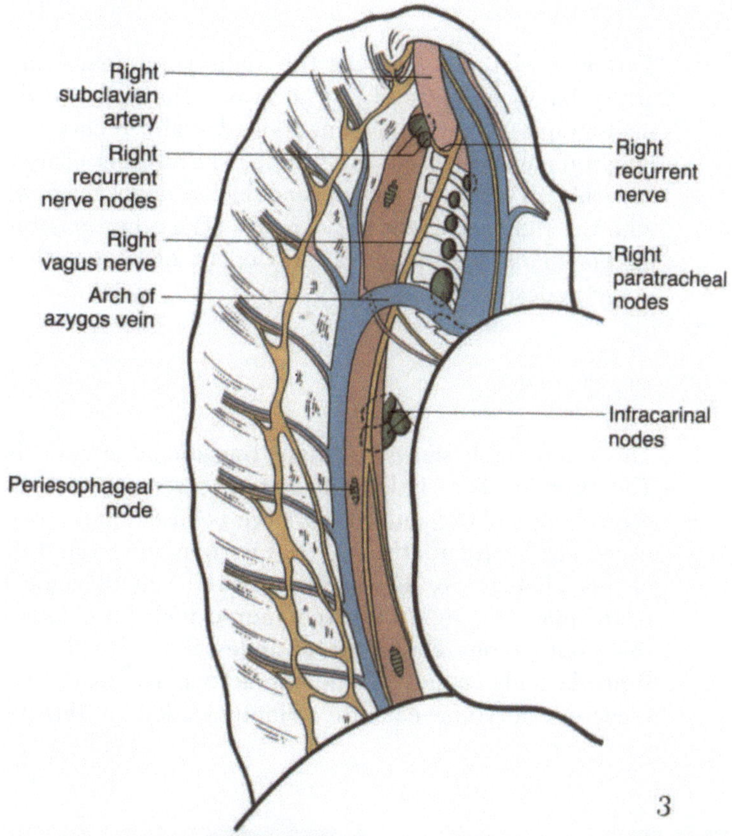
Right subclavian artery
Right recurrent nerve nodes
Right vagus nerve
Arch of azygos vein
Periesophageal node
Right recurrent nerve
Right paratracheal nodes
Infracarinal nodes

3

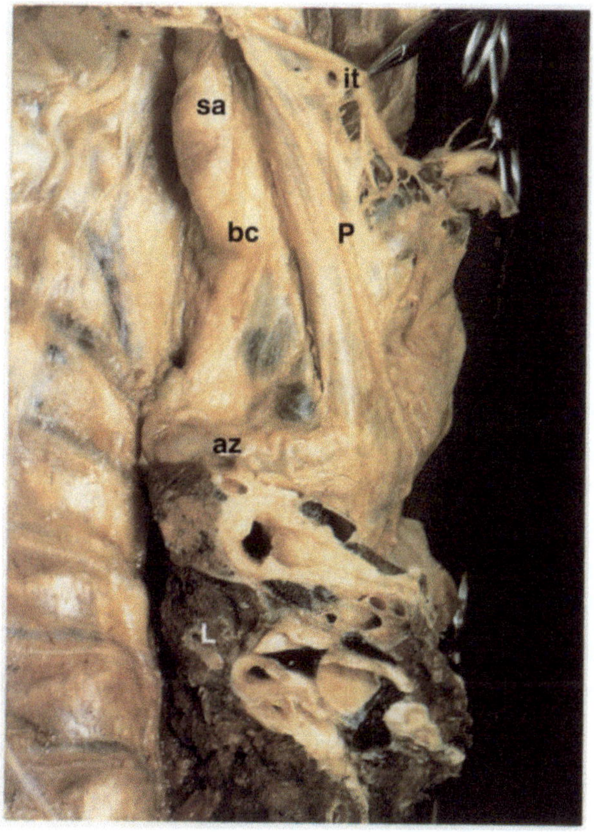
it
sa
bc
P
az
L

4

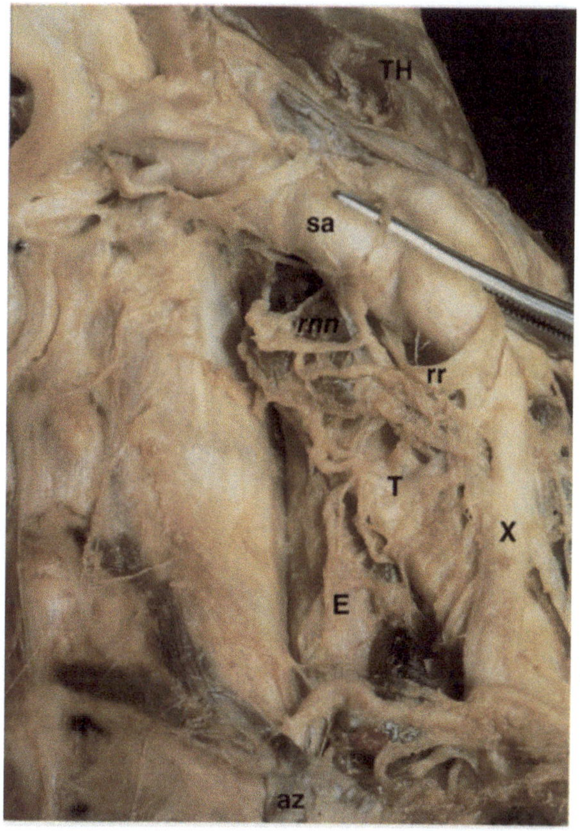
TH
sa
rnn
rr
T
X
E
az

5

Fig. 6

This is a close-up view of the right paratracheal area. The subclavian artery has been cut and moved away. The node on the *upper left* of the photograph is the above-mentioned right recurrent nerve node. Two longitudinal node chains are visible. The anterior chain bifurcates into a left superficially extending chain and into a deep chain which communicates with the right recurrent nerve nodes. The other longitudinal chain (*marked by the inlaid black papers*) also communicates with the right recurrent nerve nodes.

Fig. 7

This photograph shows the right transitional area of the neck and thorax. The thyroid gland is located in the *upper right* of this photograph. The carotid and subclavian arteries have been cut and moved away. The vagus nerve has been cut at its cervical portion and shifted to the *lower left side* of this photograph. The position of the venous angle is indicated by the *white pin*. The right recurrent nerve nodes are located in the *center* of this photograph. The group of nodes closest to the venous angle are the supraclavicular nodes. These supraclavicular nodes and those of the right recurrent nerve have a close communication via this thick lymph vessel.

Fig. 6. Connection between the right recurrent nerve nodes and the paratracheal nodes (specimen 2)

Fig. 7. Connection of the right recurrent nerve nodes and the supraclavicular nodes (specimen 2)

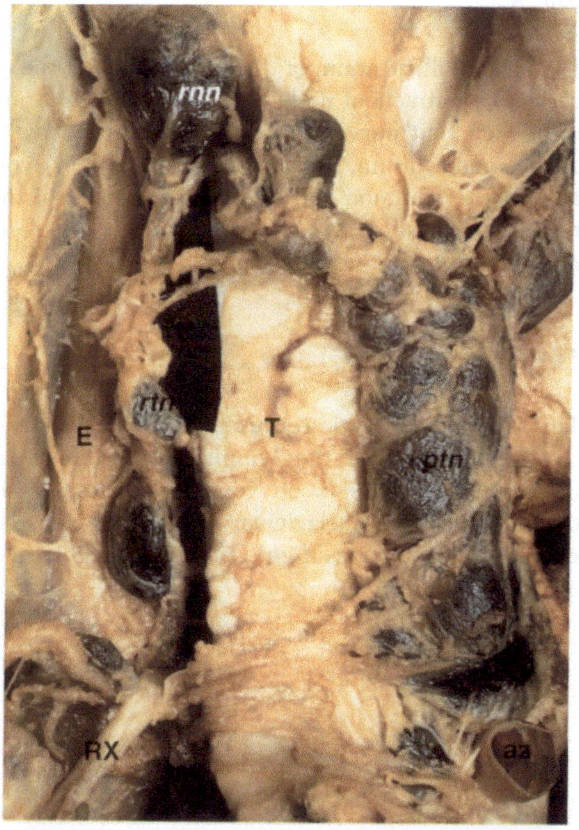

6

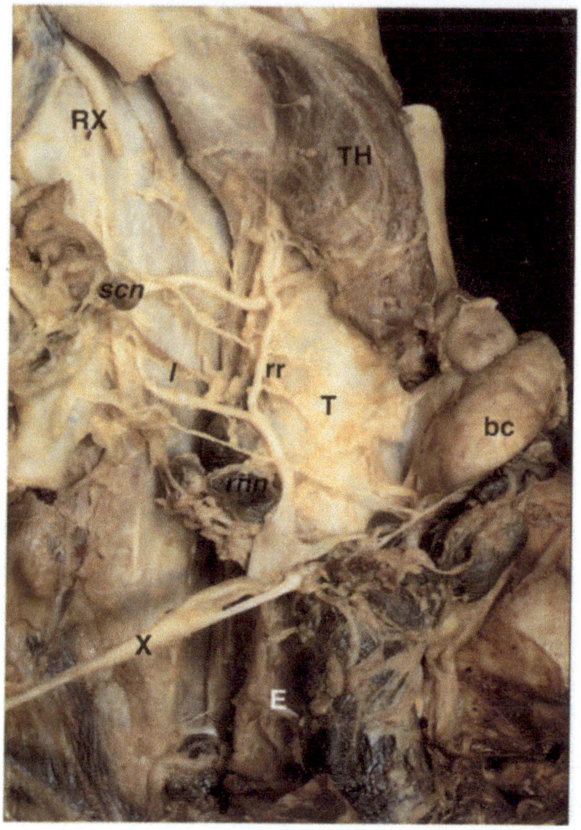

7

Fig. 8

This is an illustration of the right half of the neck (specimen 1). The cervical paratracheal nodes are seen both anterior to and posterior to the right recurrent nerve. The right recurrent nerve nodes, located at the top of the mediastinum, are connected by a thick lymph vessel which ascends behind the common carotid artery, in front of the subclavian artery, and behind the base of the internal jugular vein, reaching the supraclavicular nodes close to the venous angle.

Fig. 9

This is a diagrammatic representation of the illustration in Fig. 8, showing the relationships of the lymphatics to the surrounding structures, and in particular to the blood vessels and nerves in the lower right half of the neck. The supraclavicular and paratracheal nodes (*dark green*), right recurrent nerve nodes (*light green*), and other cervical nodes (*yellow green*) can be seen.

Fig. 8. Lymphatics of the right half of the neck (specimen 1)

Fig. 9. Diagram of the illustration in Fig. 8

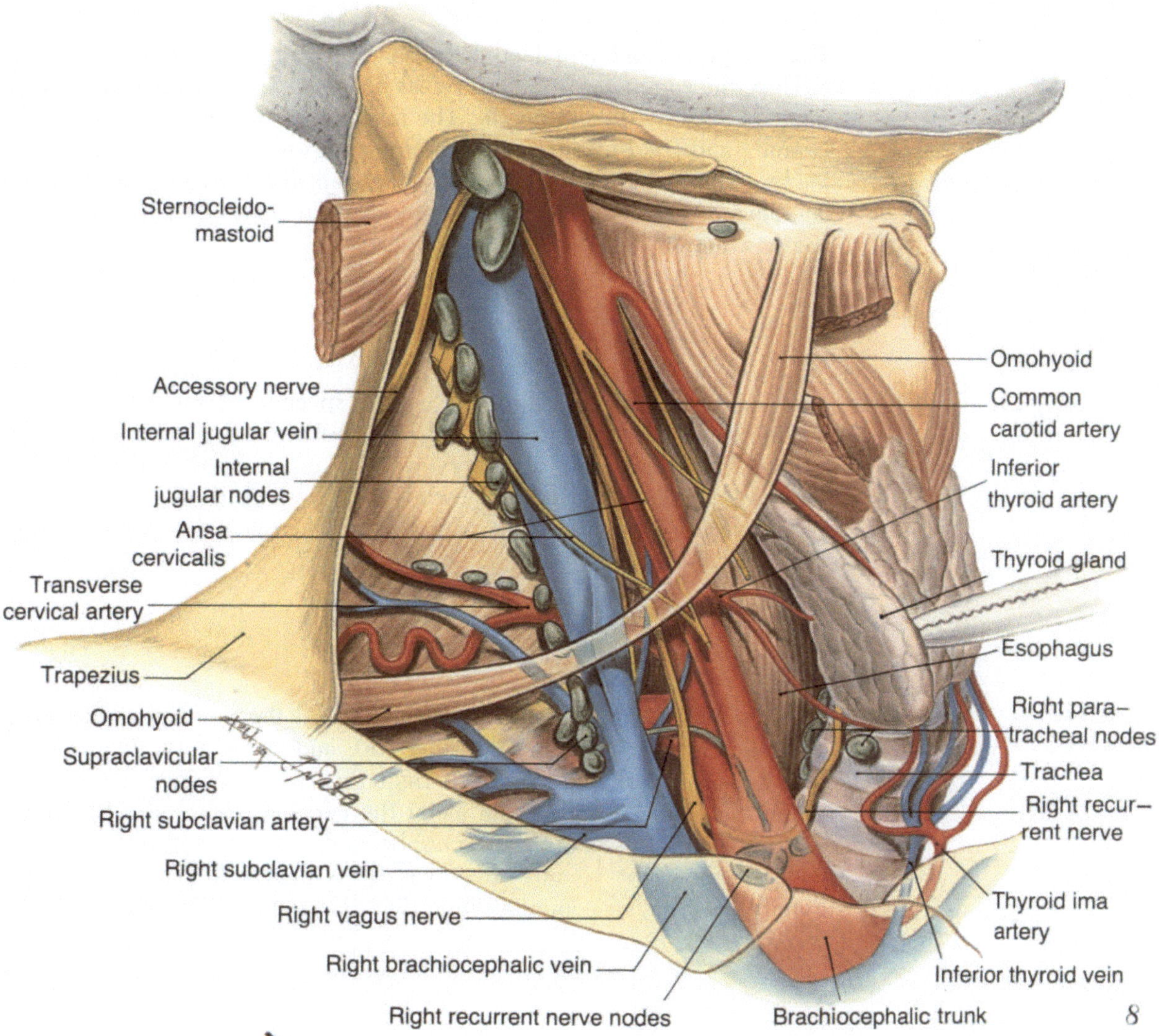

Sternocleido-
mastoid
Accessory nerve
Internal jugular vein
Internal
jugular nodes
Ansa
cervicalis
Transverse
cervical artery
Trapezius
Omohyoid
Supraclavicular
nodes
Right subclavian artery
Right subclavian vein
Right vagus nerve
Right brachiocephalic vein
Right recurrent nerve nodes
Omohyoid
Common
carotid artery
Inferior
thyroid artery
Thyroid gland
Esophagus
Right para-
tracheal nodes
Trachea
Right recur-
rent nerve
Thyroid ima
artery
Inferior thyroid vein
Brachiocephalic trunk

8

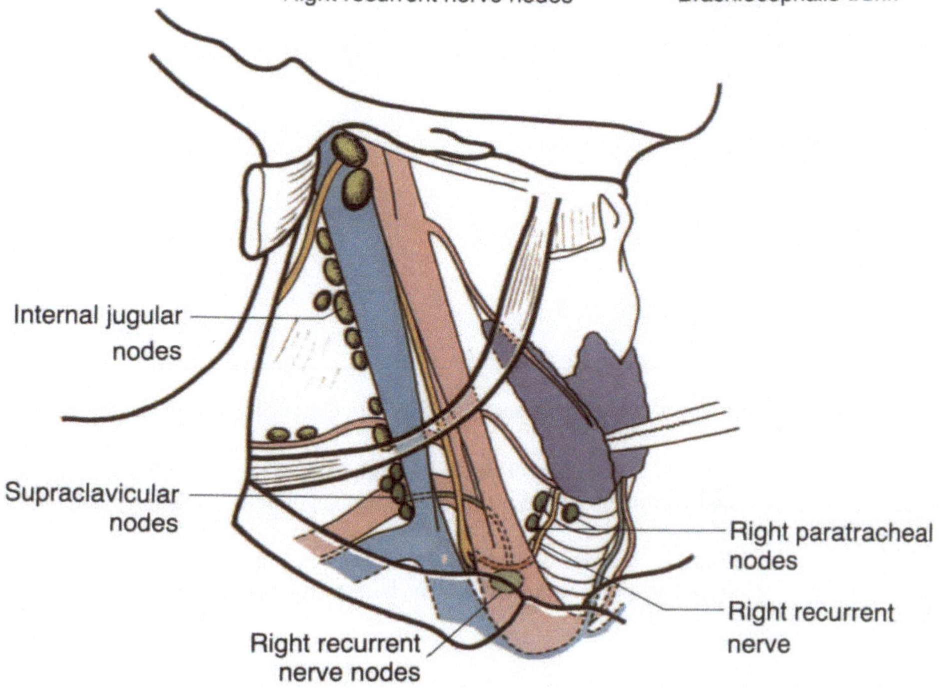

Internal jugular
nodes
Supraclavicular
nodes
Right recurrent
nerve nodes
Right paratracheal
nodes
Right recurrent
nerve

9

Fig. 10

This is a photograph of a deep dissection of the specimen illustrated in Fig. 8. Here the common carotid artery has been shifted medially and the vagus nerve is fixed by the *yellow pin*. The internal jugular vein has been shifted laterally and fixed by the *blue pin*. From the right recurrent nerve nodes (*right green arrow*), a thick lymph vessel (*left green arrow*) makes an arch lateralwards and extends to the supraclavicular nodes close to the venous angle.

Fig. 11

This photograph of the right neck area of specimen 3 was taken after the common carotid artery (*lowest red pin*) and internal jugular vein (*upper blue pin*) were cut. The vagus nerve was left intact (*upper yellow pin*). The three segments of the carotid sheath which correspond to the artery, vein, and nerve were longitudinally cut in the middle, opened, and pinned using the respective colored pins.

Fig. 12

This is a photograph of the same specimen after the common carotid artery (*red pin*), internal jugular vein (*blue pin*), and vagus nerve (*lower yellow pin*) were cut and reflected. The right recurrent laryngeal nerve (*upper yellow pin*) and its node chain along the cervical trachea can be seen. A communicating vessel (*green arrows*) between one of the chain nodes and a node of the right venous angle has been dissected. This vessel is an example of a higher communication than that seen in Figs. 6 and 10.

Fig. 10. Deeper dissection of the specimen illustrated in Fig. 8

Fig. 11. Dissection of the lower right half of the neck after the internal jugular vein and common carotid artery were cut (specimen 3, male)

Fig. 12. Deeper dissection of specimen in Fig. 11, showing a communicating vessel from the right recurrent nerve chain to the jugular chain

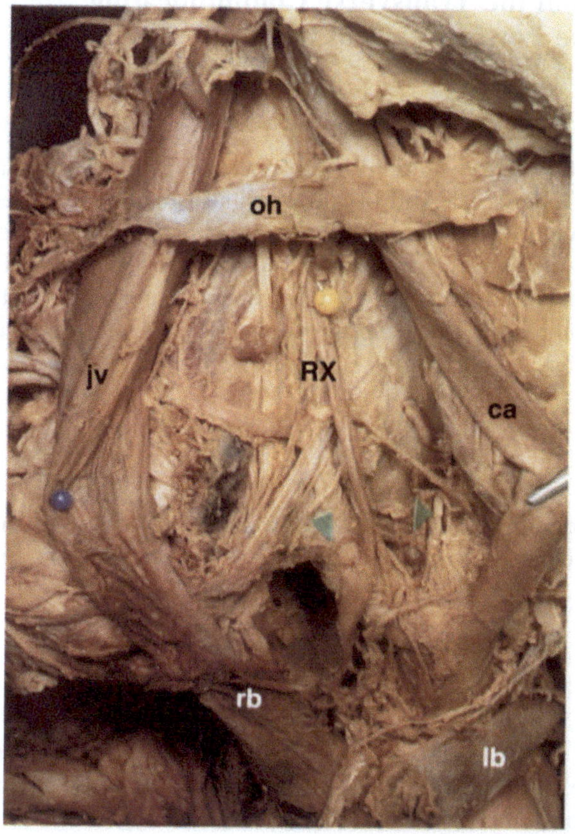

10

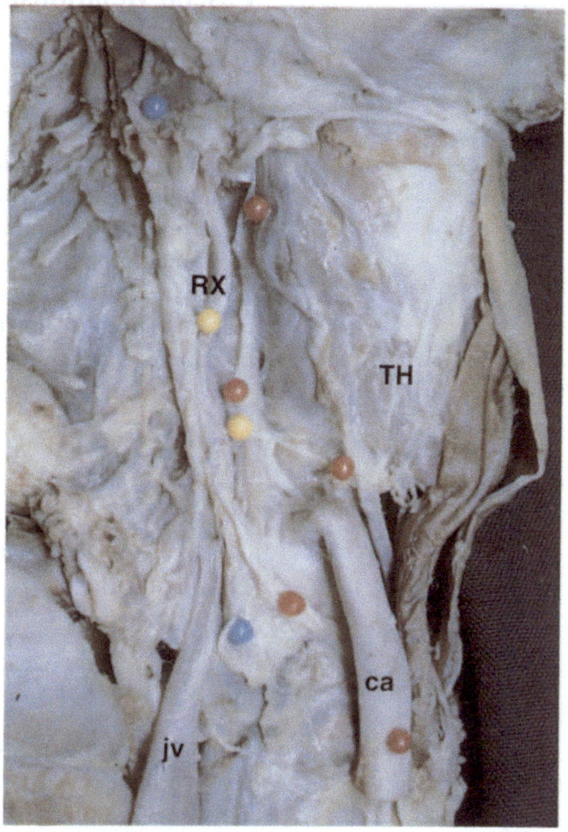

11

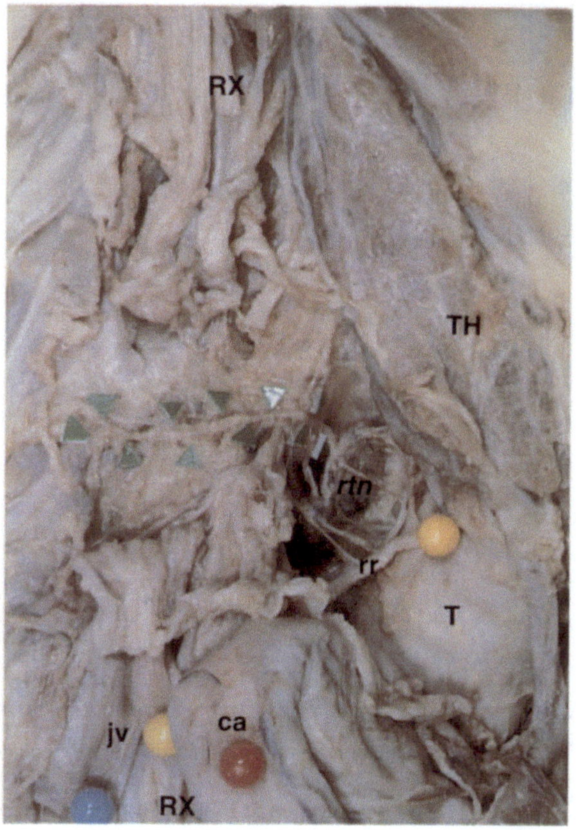

12

Fascial Relationship of the Transverse Communicating Lymph Vessels in the Lower Cervical Region

Fig. 13

This is a diagrammatic transverse section of the neck, showing the fascial arrangement. Note that the visceral and prevertebral fasciae are connected by strong coalescent connective tissues via the carotid sheath; these tissues are considered to be septa which enclose the retropharyngeal space. In order for the transverse lymph vessel from the recurrent nerve chain to exit this space, it must pierce the thick septum connecting the prevertebral fascia and common carotid sheath (*arrow*).

Figs. 14–16

This is a series of photographs showing the actual dissection view of the fascial connection between the visceral and prevertebral fasciae via the carotid sheath.

Fig. 14

This is a photograph of a transverse section of specimen 4. The larynx has been pulled toward the front allowing view of the loose connective tissue between the pharynx and the cervical vertebral column.

Fig. 13. Transverse section diagram of the neck muscles and fasciae

Fig. 14. Loose connective tissue in the retropharyngeal space (specimen 4, male)

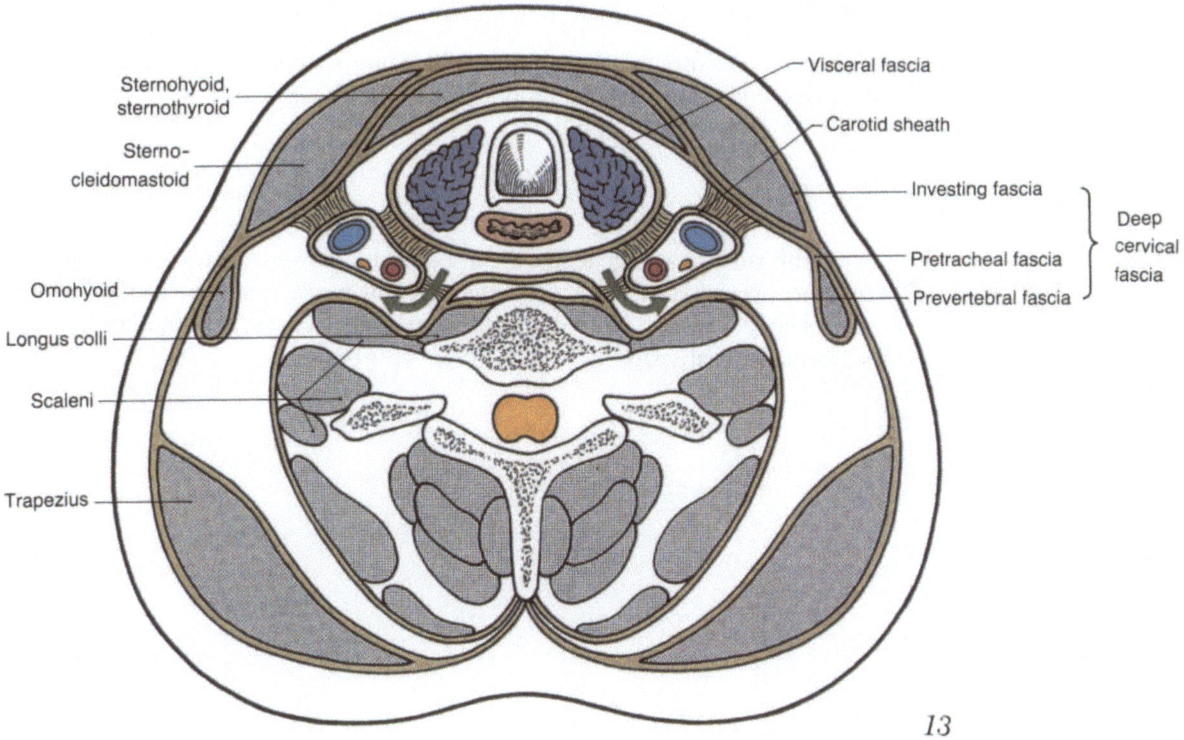
Sternohyoid, sternothyroid
Sterno-cleidomastoid
Omohyoid
Longus colli
Scaleni
Trapezius
Visceral fascia
Carotid sheath
Investing fascia
Pretracheal fascia
Prevertebral fascia
Deep cervical fascia

13

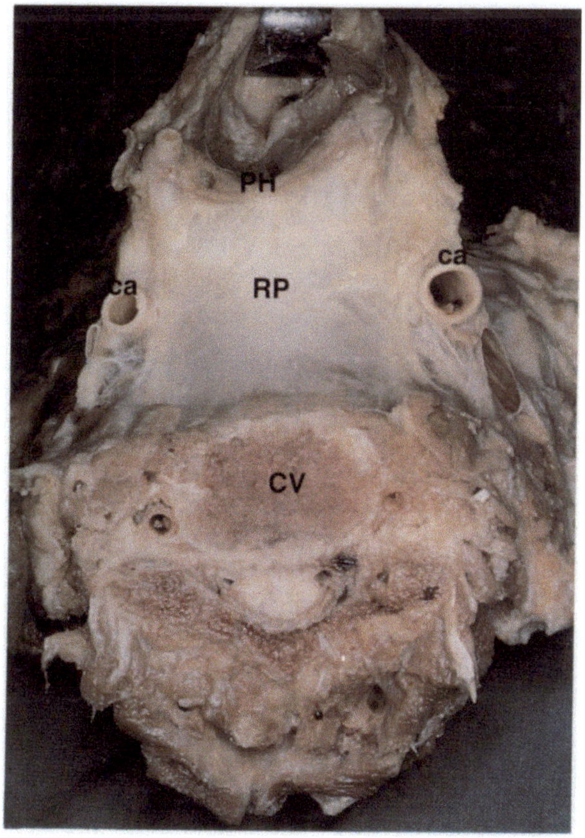
PH
ca
RP
ca
CV

14

Fig. 15

This is a subsequent photograph of the same specimen. The loose connective tissue is being cleared away by use of forceps.

Fig. 16

Now the retropharyngeal space can be easily viewed. Thick septa connecting the prevertebral fascia and the visceral fascia, via the carotid sheath, are clearly visible. Note that these septa serve as a boundary for the retropharyngeal space.

Fig. 15. Gentle clearing of the retropharyngeal space (specimen 4)

Fig. 16. Thick septa encircling the retropharyngeal space (specimen 4)

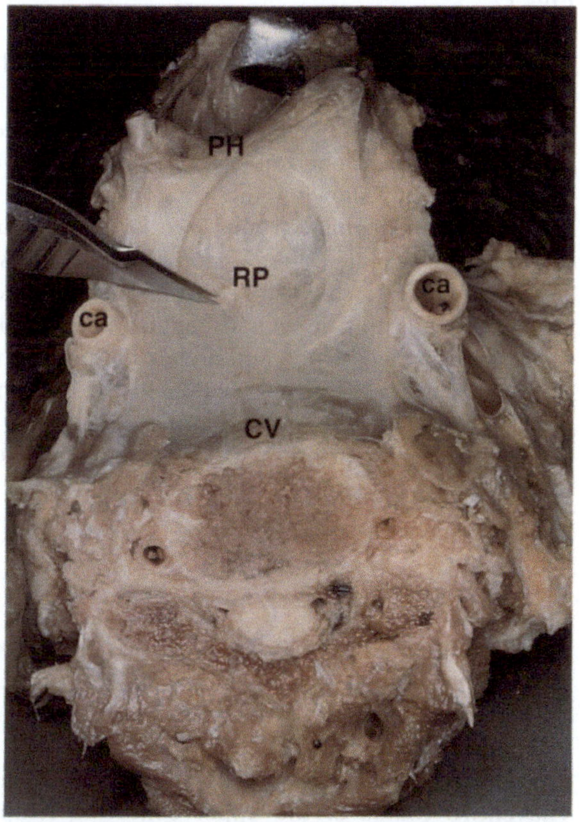

15

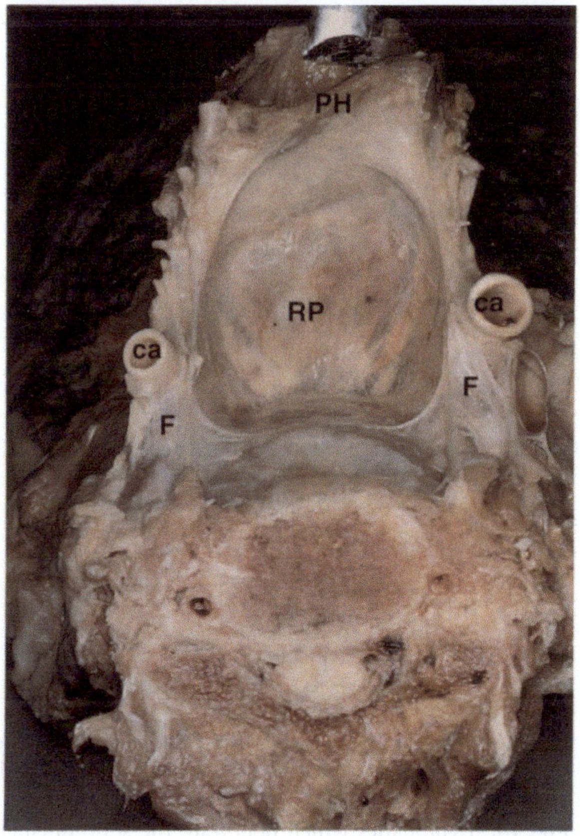

16

Fig. 17

This is a photograph of the right cervicomediastinal transitional area (specimen 5) seen from the front and side. The cervical viscera with the carotid sheath have been pulled anteriorly to show the lower part of the thick septum mentioned above. The right brachiocephalic vein has been shifted laterally, showing the right paratracheal area between this vein and the brachiocephalic trunk. Two large collecting lymph vessels leading to the venous angle can be seen (*green arrows*). The lower, and more superficial vessel which arises from the paratracheal nodes at the level of the brachiocephalic venous angle ascends along the lateral flank of the brachiocephalic trunk and arches behind the vagus nerve and over the subclavian artery, with no relation to the septum mentioned above. The upper vessel, on the other hand, pierces the septum before reaching the venous angle. In order to show the complete course of this vessel, it was necessary to remove the thick septum.

Fig. 18

This photograph shows a deeper dissection of Fig. 17, after removing the septum and cutting the brachiocephalic and common carotid arteries and the cervical segment of the vagus nerve, leaving the thoracic segment intact. The subclavian artery has been reflected laterally to allow view of the right recurrent laryngeal nerve and its accompanying lymphatic chain. The entire course of the upper collecting lymph vessel (*green arrows*), which arises from a node of the recurrent nerve chain, is clearly visible. Note that this vessel runs deep to the recurrent laryngeal nerve.

Fig. 17. Dissection showing the relationship between the lymphatics and the septum connecting the carotid sheath and prevertebral fascia (specimen 5, male)

Fig. 18. Lymphatics which penetrate the septum (specimen 5)

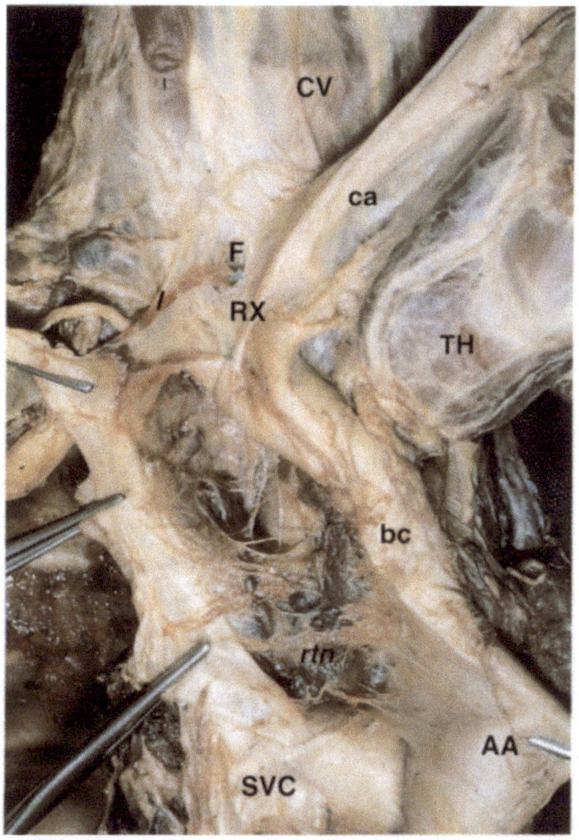

17

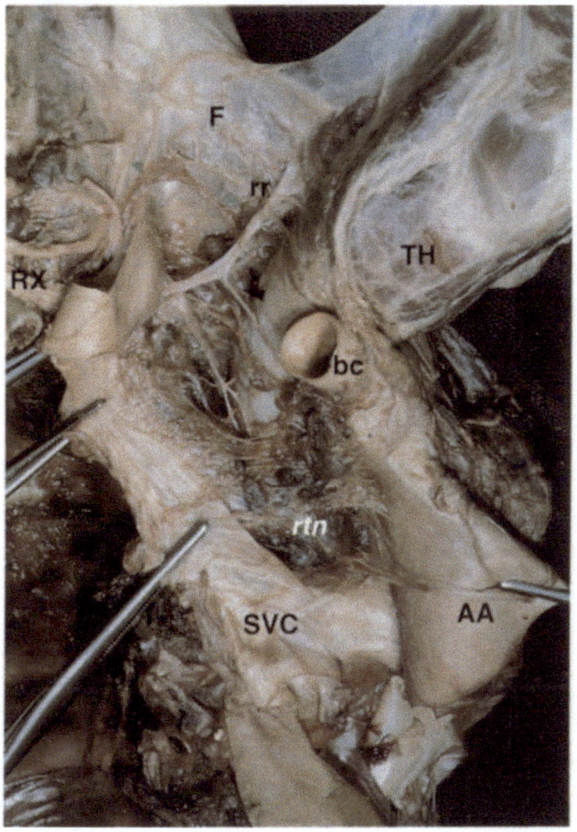

18

Right Superficial Pathway

Figs. 19–25

This is a series of photographs showing the dissection of a right superficial pathway leading from the paratracheal nodes to the venous angle (specimen 6, male).

Fig. 19

This photograph shows the lymphatics of the area surrounding the tracheal bifurcation (refer to Fig. 1). The heart, pericardium, great vessels, and the anterior layer of the pulmonary ligaments have been removed. The aortic arch with the pulmonary artery has been shifted to the left (*right margin* of the photograph), allowing clear view of the well-developed nodes.

Fig. 20

This is a closer view of Fig. 19. In the pulmonary ligaments, several lymph vessels are seen along the esophagus, which ascend medial to the section of the right inferior pulmonary vein and lateral to that of the left corresponding vein (*two lowest right arrows*) and reach the infracarinal nodes directly or via the bronchopulmonary nodes. Numerous communicating vessels can be seen from the bronchopulmonary and infracarinal nodes to the right and left tracheobronchial nodes, with more extending to the right.

Fig. 19. Dissection of the right superficial pathway from the paratracheal nodes (specimen 6, male)

Fig. 20. Communication between the infracarinal nodes and the right paratracheal nodes (specimen 6)

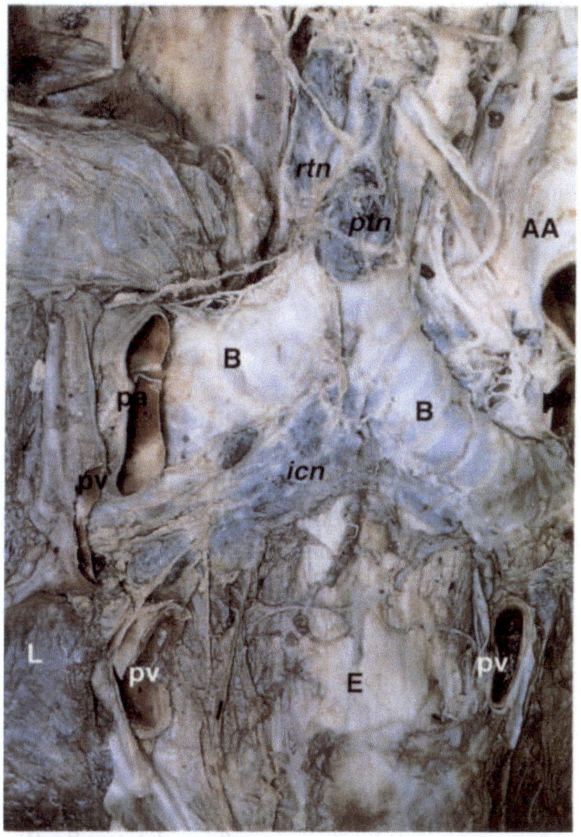

19

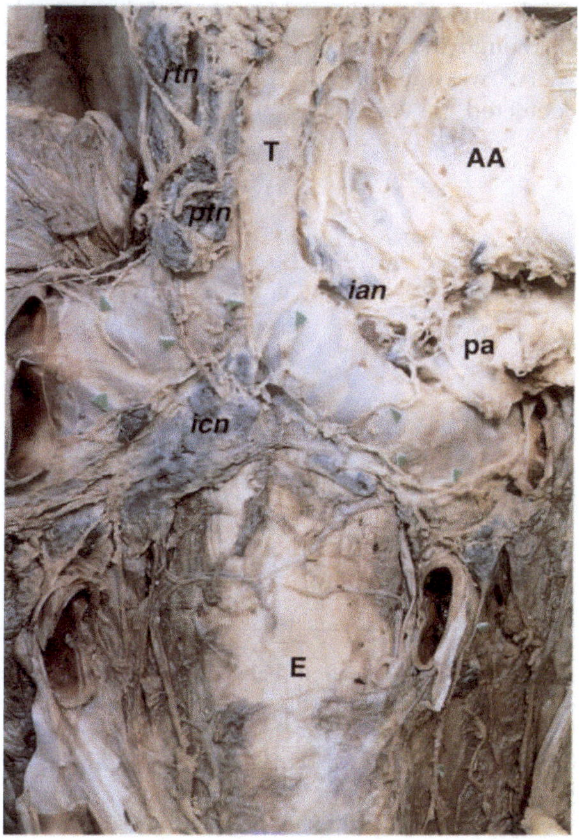

20

Fig. 21

This photograph shows the cervicomediastinal area of Fig. 20. The well-developed nodes *near the top of the previous two figures* are shown *here* at the *lower right* of the figure and just below the right brachiocephalic vein (*blue pin*). These are two adjacent right paratracheal nodes, one in front of and one to the side of the trachea (pre- and paratracheal nodes). Note the numerous superficial vessels which run anterior to the right brachiocephalic vein from these nodes to the sapraclavicular nodes.

Fig. 22

In this deeper dissection than that shown in the previous figure, the brachiocephalic trunk and an autonomic cardiac nerve (formed by the union of branches of the vagus and sympathetic nerve) have been reflected cranially and pinned (*red and yellow pins*) allowing view of the right recurrent laryngeal nerve and its surrounding connective tissue. The *green pin* indicates a lymph vessel from the top of the above-mentioned nodes to the left venous angle along the left brachiocephalic vein. This vessel is a superficial communication between the right and left pathways.

Fig. 23

This is a further dissection of the specimen in the previous figure. The common carotid artery is reflected laterally (Pean's forceps) allowing view of the area along the right margin of the thyroid gland and cervical trachea. Although the connective tissue surrounding the right recurrent laryngeal nerve was meticulously removed, no deep lymph nodes or vessels were detected.

Fig. 21. Communication between the right paratracheal nodes and the right venous angle (specimen 6).

Fig. 22. The right recurrent laryngeal nerve and its surrounding tissues (specimen 6)

Fig. 23. Gentle clearing of the right recurrent nerve area (specimen 6)

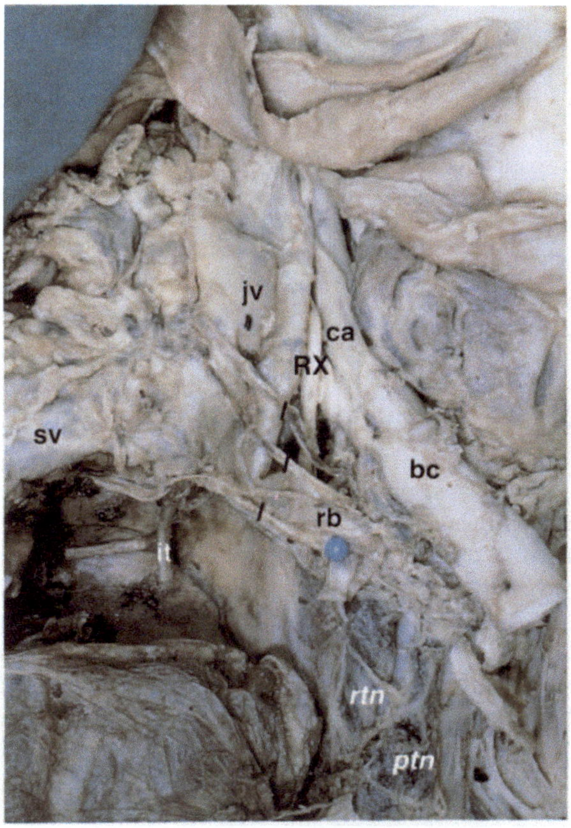
sv
jv
ca
RX
sv
bc
rb
rtn
ptn

21

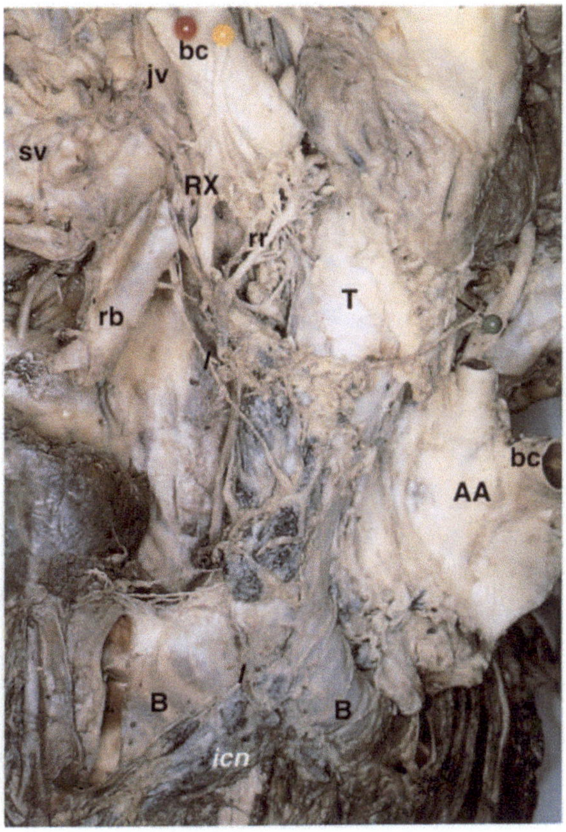
bc
jv
sv
RX
rr
T
rb
bc
AA
B
B
icn

22

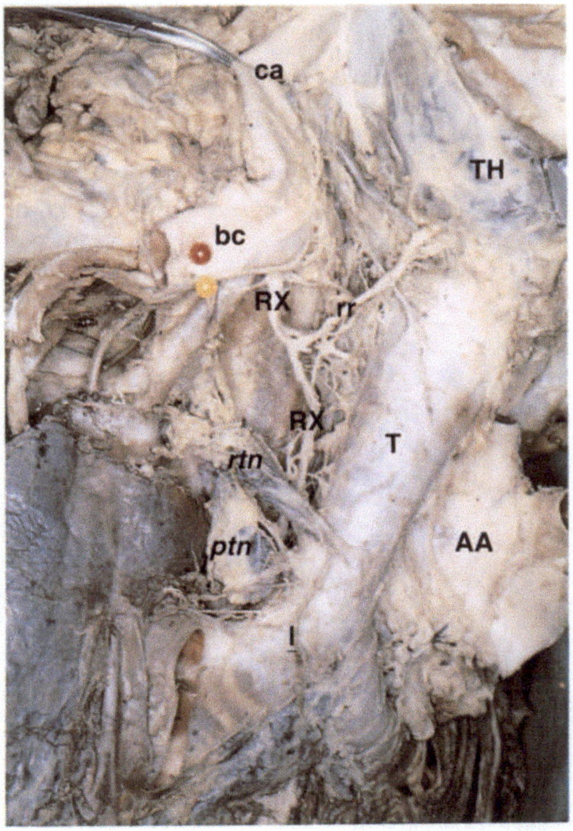
ca
TH
bc
RX
rr
RX
T
rtn
ptn
AA

23

Fig. 24

In this photograph, the right brachiocephalic vessels are cranially reflected and pinned, showing the posterior area of the venous angle just medial to the scalenus anterior. In addition to the above-mentioned superficial collecting vessels which pass over the brachiocephalic vein (Fig. 21), another long lymph vessel, which passes obliquely beneath this vein, can be seen.

Fig. 25

This photograph is the same view of the specimen in Fig. 24; however, *black paper* has been inlaid to show the course of the above-mentioned long lymph vessel. Note that this vessel reaches the posterior area of the venous angle.

In this specimen the right pathway was composed of only superficial collecing vessels which ascended over and beneath the brachiocephalic vein. Based on the right pathway findings in the above six specimens, the course leading to the right venous angle can be divided into the following three types according to the positional relationship with the right brachiocephalic vessels: (1) Deep pathway ascending behind the brachiocephalic trunk, (2) posterior superficial pathway ascending beneath the right brachiocephalic vein, and (3) anterior superficial pathway ascending over the right brachiocephalic vein.

Fig. 24. A superficial connection along the brachiocephalic trunk (specimen 6)

Fig. 25. Connecting lymph vessel (seen in Fig. 24) which reaches the right venous angle (specimen 6)

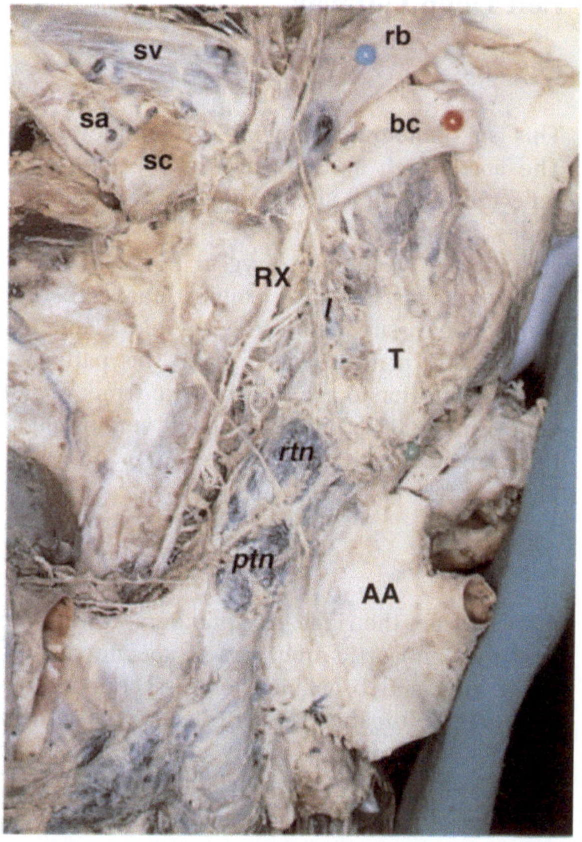

24

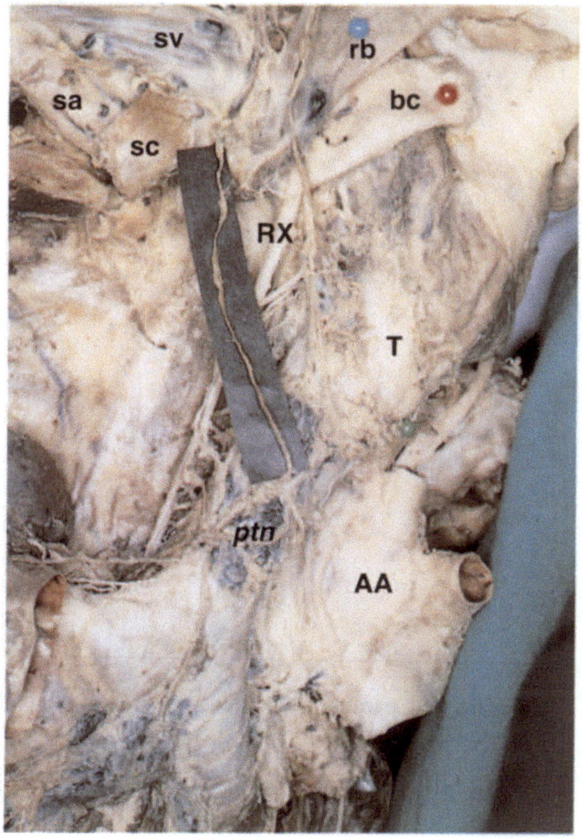

25

Left Deep Pathway

Fig. 26

This is a photograph showing the complete left deep course of specimen 6. The aortic arch and brachiocephalic trunk (*red pin*) have been strongly reflected to the left after removal of the major vessels. From the infracarinal nodes (*lowest green arrow*) this pathway ascends between the trachea and the left recurrent larygeal nerve and anterior to the left margin of the esophagus. A small connecting vessel obliquely descends from the connective tissue below the thyroid gland to join this pathway. At the inferior end of the thyroid gland this pathway begins to arch to the left, passing anterior to the sympathetic trunk (*yellow arrow*) and the vertebral vein. Finally, it drains into the left venous angle close to the termination of the thoracic duct.

Fig. 27

Viewed from the right, this photograph shows the left margin of the trachea and the infraaortic arch area of specimen 1, after strong reflection of the right lung and bronchus (for dissection details, see Fig. 28). The aortic arch, the bifurcation of the trachea, the left bronchus, the left pulmonary arteries, the bronchial artery, and the left recurrent laryngeal nerve are all clearly visible. The infraaortic arch nodes and the left paratracheal nodes also can be seen.

Fig. 26. Left paratracheal nodes and the deep pathway to the left venous angle (specimen 6)

Fig. 27. Dissection of the left paratracheal nodes viewed from the right (specimen 1).

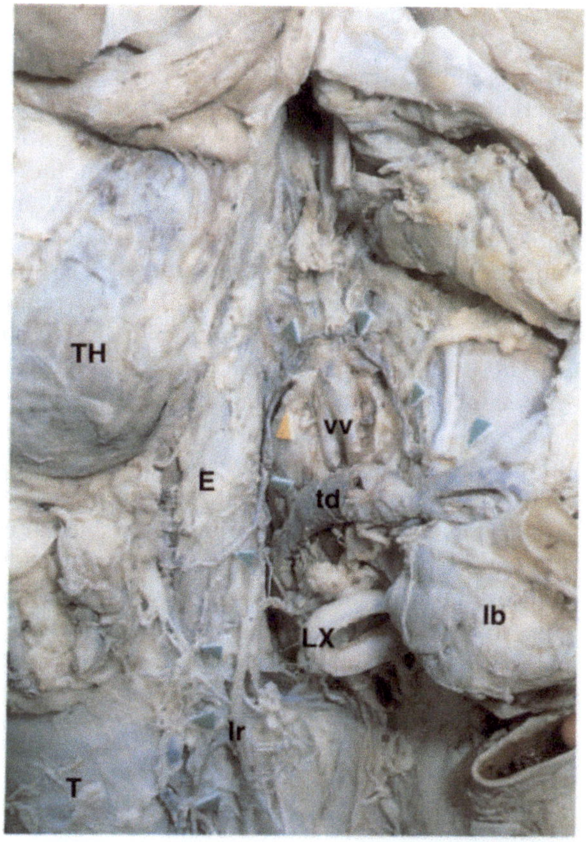

26

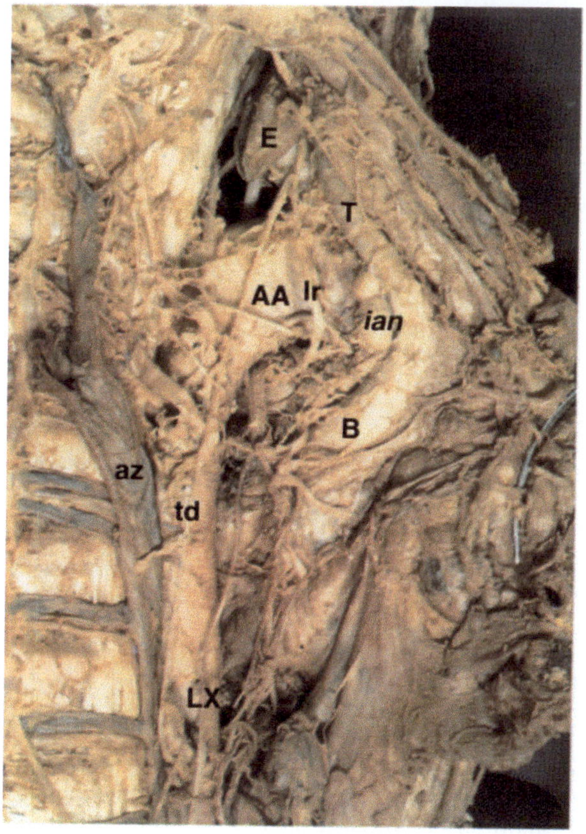

27

Fig. 28

This is an illustration of the left deep lymph pathway based on the dissection of specimen 1, viewed from the right. The bronchial artery from the highest intercostal artery of aortic origin has been cut, and the right lung and bronchus have been pulled to the front and to the left. The esophagus has been cut at the top of the thorax and shifted to the right after cutting the vagus nerve just under the points of origin of the pulmonary branches. Thus, the reflected posterior surface of the trachea, and the right and left bronchi and the surrounding structures are visible. The distal half of the aortic arch and the left subclavian artery, as well as the descending aorta and the thoracic duct, can be seen. Note the space between the aortic arch and the left bronchus. In the lower half of this space the left pulmonary artery is partially visible. The arch of the aorta and the left pulmonary artery are connected by the ligamentum arteriosum. The left vagus nerve can first be seen inferior to the aortic arch. The left recurrent laryngeal nerve from the vagus nerve winds around the ligamentum arteriosum, ascending along the left margin of the trachea. Regarding the lymph nodes of the left side of the trachea and esophagus, the left paratracheal nodes form an ascending lymphatic chain, which begins with the infraaortic arch nodes and runs medial to the recurrent laryngeal nerve, and between this nerve and the trachea. In addition to the many paratracheal nodes, there is a lymph vessel from the esophagus which runs along an esophageal branch of the aorta and drains into the thoracic duct.

Fig. 28. Left paratracheal nodes viewed from the right based on the dissection shown in Fig. 27 (specimen 1)

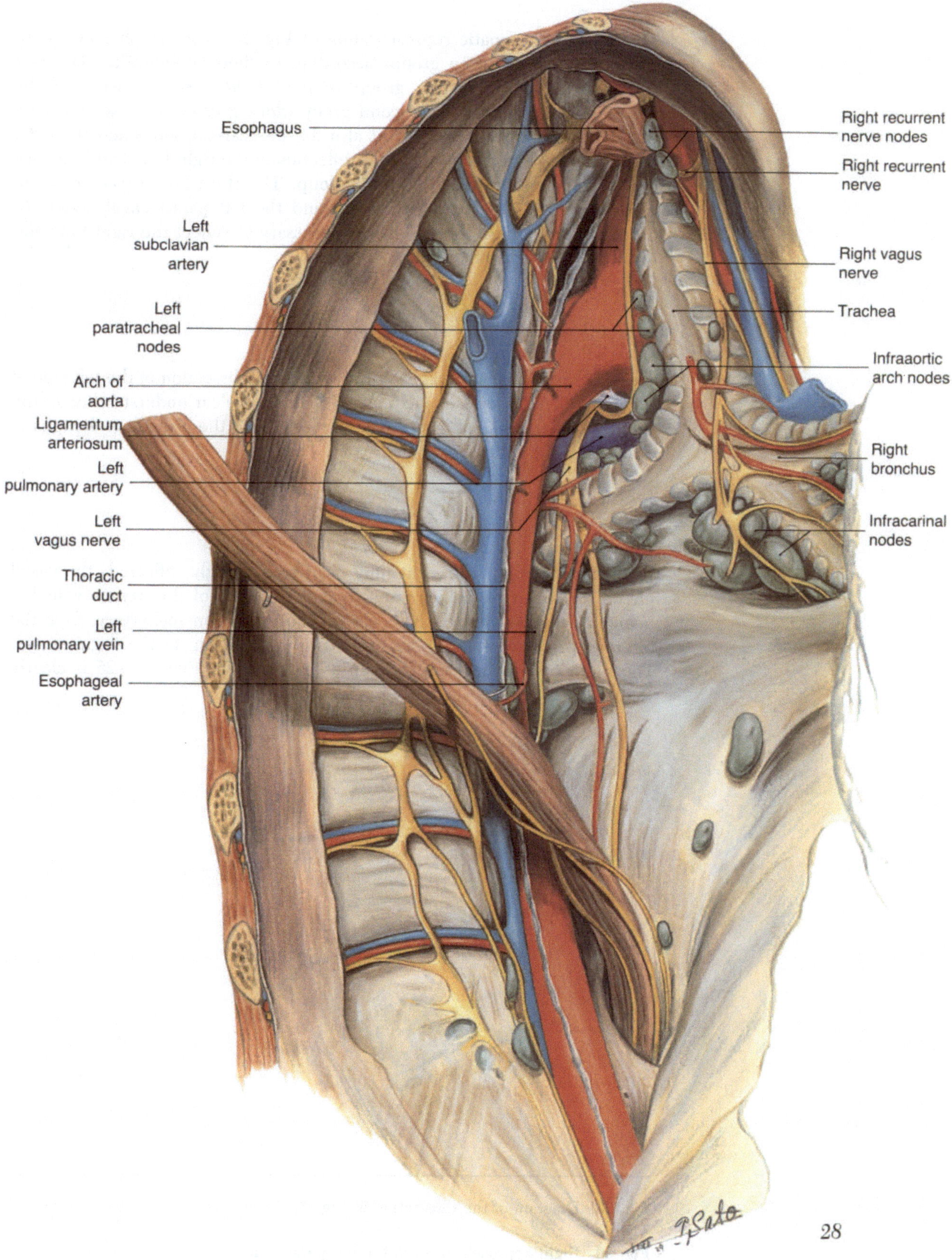
Esophagus
Left subclavian artery
Left paratracheal nodes
Arch of aorta
Ligamentum arteriosum
Left pulmonary artery
Left vagus nerve
Thoracic duct
Left pulmonary vein
Esophageal artery
Right recurrent nerve nodes
Right recurrent nerve
Right vagus nerve
Trachea
Infraaortic arch nodes
Right bronchus
Infracarinal nodes

28

Fig. 29

This is a diagrammatic representation of Fig. 27. The lymph nodes here are classified into three groups according to their accessibility. The first and most easily accessible group (*striped light green*) consists of the periesophageal nodes. The second group (*dark green*) comprises not only the right paratracheal nodes, but also the infracarinal nodes and the right recurrent nerve nodes. Only slight reflection of the right lung and bronchus is necessary to approach this right group. The third group (*yellow-green*) consists of the infraaortic arch nodes and the left paratracheal nodes. In order to access these left nodes it is necessary to reflect the right lung and bronchus to the front.

Figs. 30–35

This is a series of photographs of the sequential dissection of the left side of specimen 6. This series is included to facilitate a clear understanding of the step-by-step dissection procedure for reaching the left deep lymphatic chain from the right.

Fig. 30

In this photograph, the right lung has been strongly reflected, the costal pleura has been partially removed, and the arch of the azygos vein has been cut to allow view of the course of the right bronchial artery, from the intercostal artery of aortic origin, and its relationship to the vagus nerve. The same paratracheal node group as that seen in Figs. 19–25 is clearly visible.

Fig. 29. Diagram of the illustration in Fig. 27

Fig. 30. Right approach to reach the left paratracheal nodes (specimen 6)

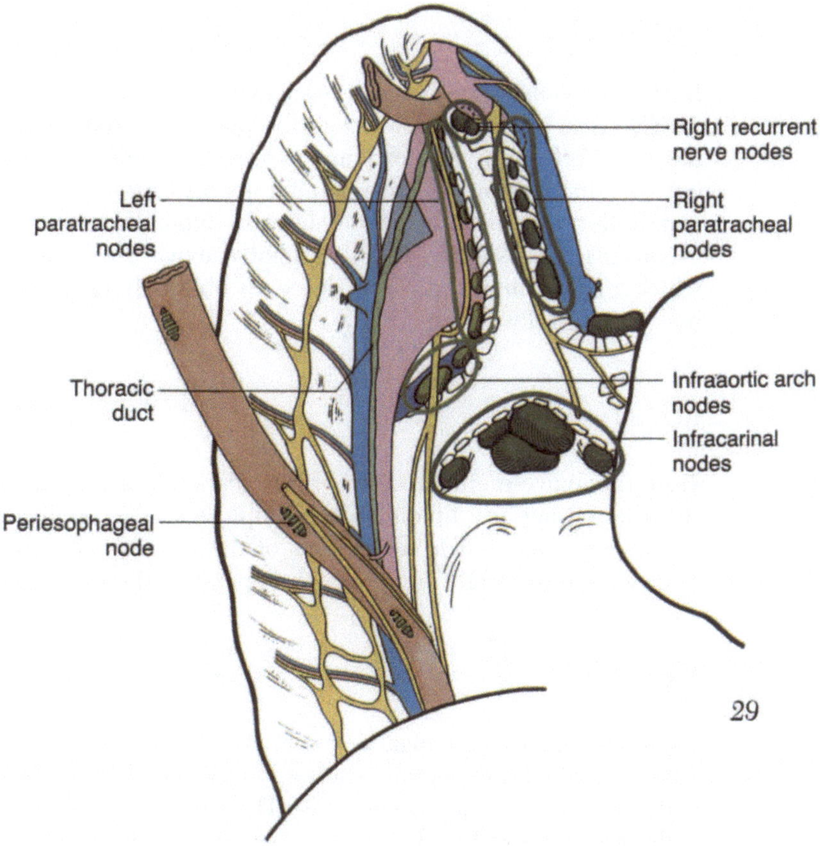
Right recurrent nerve nodes
Left paratracheal nodes
Right paratracheal nodes
Thoracic duct
Infraaortic arch nodes
Infracarinal nodes
Periesophageal node

29

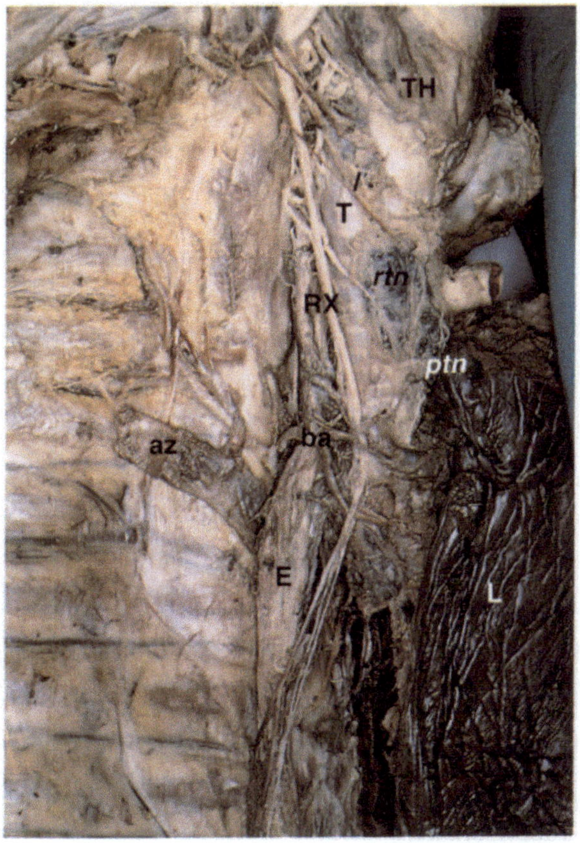
TH
T
rtn
RX
ptn
az
ba
E
L

30

Fig. 31

In this close-up view, the right bronchial artery has been cut before its first division into branches posterior to the pulmonary root. The inferior branch has been reflected anteriorly and pinned (*red pin*). The vagus nerve was cut after its division into two parallel nerves (*yellow pins*) which pass over and under the inferior arterial branch. Note the node attached to the esophagus (periesophageal node) slightly above the bronchial artery, from which a communicating vessel descends obliquely to reach the infracarinal nodes (*near the red pin*).

Fig. 32

Here the above-mentioned periesophageal node is pulled with a forceps to show its strong attachment to the esophagus. The trachea and esophagus have been separated to allow view of the infracarinal nodes, as well as a branch from the vagus nerve which innervates the esophagus.

Fig. 33

The above-mentioned vagal nerve branch has been cut, facilitating a clear view of the oblique communicating lymph vessel between the infracarinal nodes and the periesophageal node. The connective tissue surrounding this node was removed and now the lymph vessels which originate at the esophageal wall can be seen.

Fig. 31. Right vagus nerve and bronchial artery and their relationship to the surrounding organs (specimen 6)

Fig. 32. Periesophageal node strongly attached to the esophagus (specimen 6)

Fig. 33. Communication between the periesophageal node and the infracarinal nodes (specimen 6)

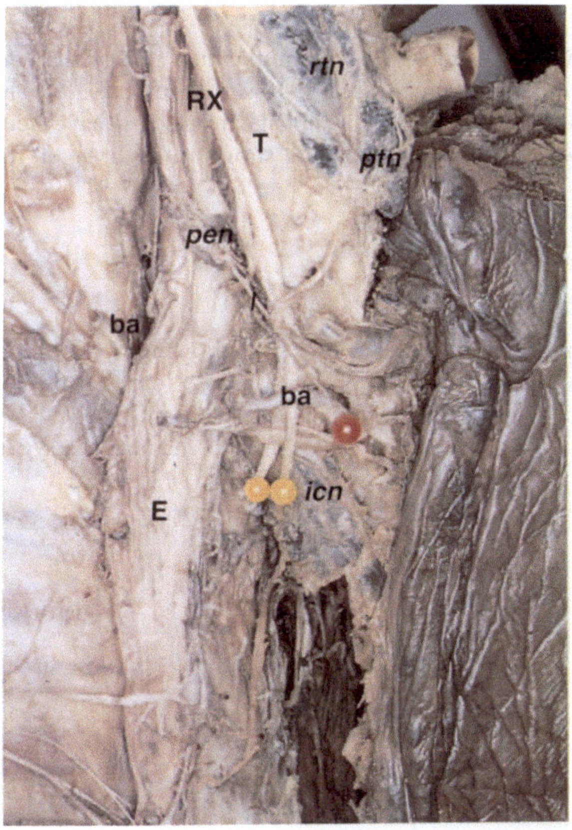

31

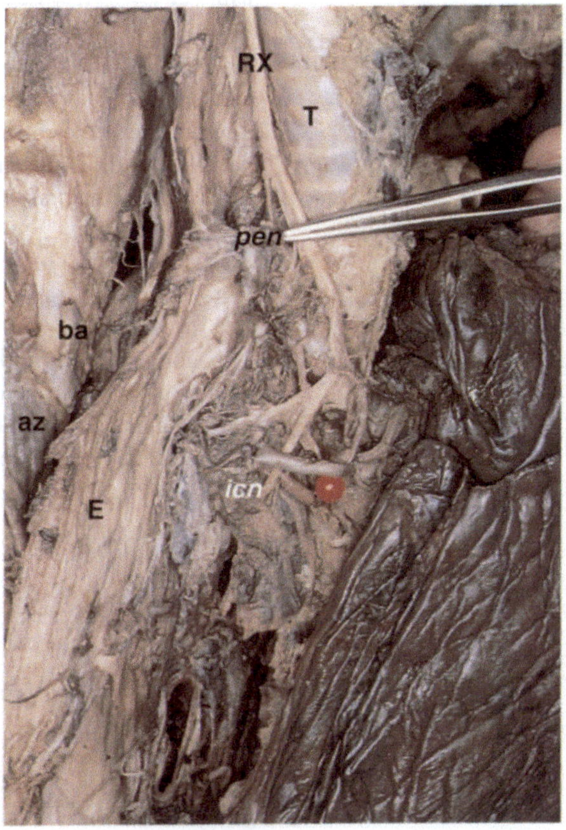

32

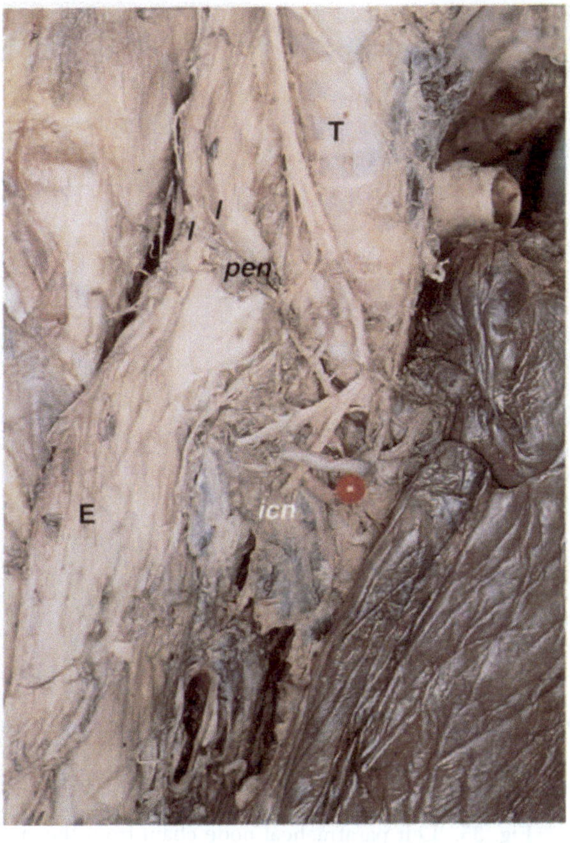

33

Fig. 34

Here, the above-mentioned esophageal communication has been cut and the periesophageal node has been shifted toward the ***red pin***. The dense transverse muscular bundles (tracheoesophageal muscles) have been detached, allowing view of the lightly shaded supracarinal area of the anterior esophagus.

Fig. 35

The above-mentioned muscular bundles were cut, allowing view of the ascending structures along the left margin of the trachea. Note that the left paratracheal node chain, originating from the infraaortic arch nodes, lies between the left margin of the trachea and the left recurrent larygeal nerve.

Fig. 34. Separation of the trachea and esophagus (specimen 6)

Fig. 35. Left paratracheal node chain from the right (specimen 6)

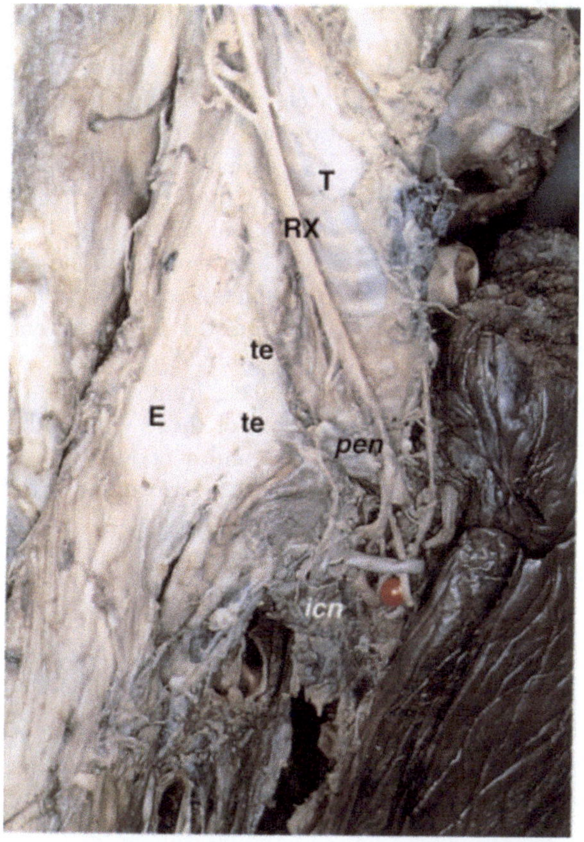

34

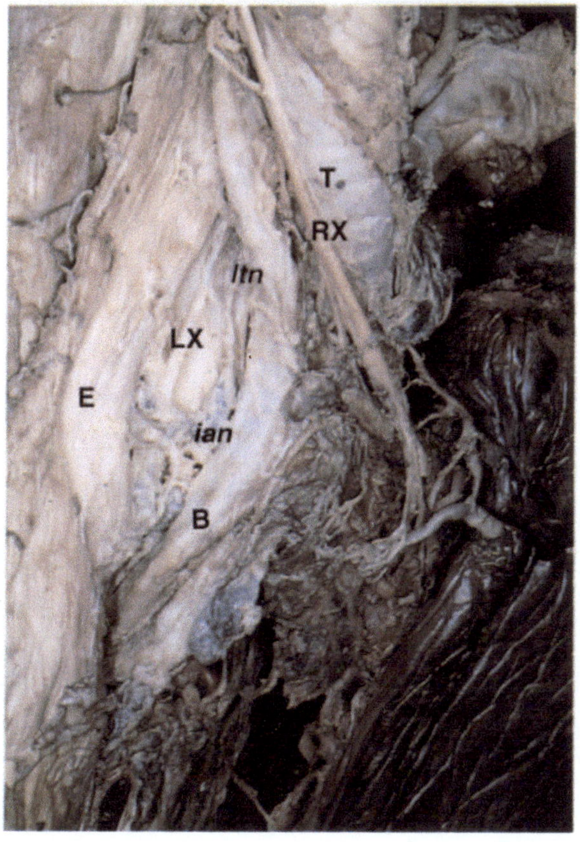

35

Fig. 36

This is a photograph of a deep dissection of another specimen (specimen 7), included to provide a clear understanding of the left paratracheal region from a different point of view. Here the lymphatics surrounding the upper thoracic portion of the esophagus, as viewed from the front and slightly from the left, are shown. At the *upper left* the thyroid gland can be seen and at the *bottom*, the pulmonary trunk. The proximal portion of the aortic arch has been cut and removed. The trachea and the left margin of the esophagus are visible. The *red arrow* indicates the ligamentum arteriosum, the *yellow arrow* the left recurrent laryngeal nerve, and the *green arrows* the thoracic duct. Small *black papers* have been inserted, making several lymph vessels from the well-developed left paratracheal nodes clearly visible. Some of these vessels drain into the thoracic duct. The infraaortic arch nodes and a node of the ligamentum arteriosum are separated by the ligamentum arteriosum (*red arrow*).

Fig. 37

This is a magnification of part of Fig. 36. The *green arrow* indicates the lymph vessels which originate from the esophageal wall and drain into the thoracic duct.

Fig. 36. Left paratracheal node chain, its communication, and its relationship to the thoracic duct (specimen 7, male)

Fig. 37. Esophageal lymph vessel which drains into the thoracic duct (specimen 7)

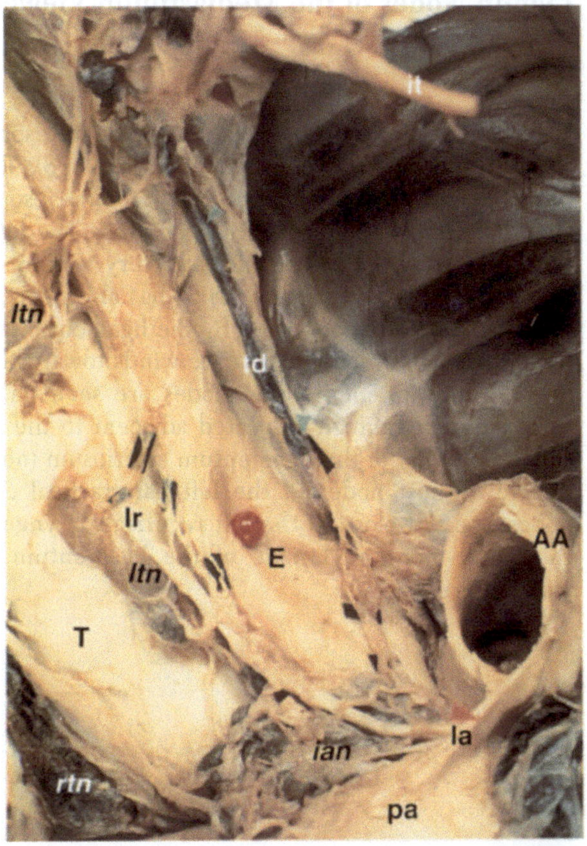

36

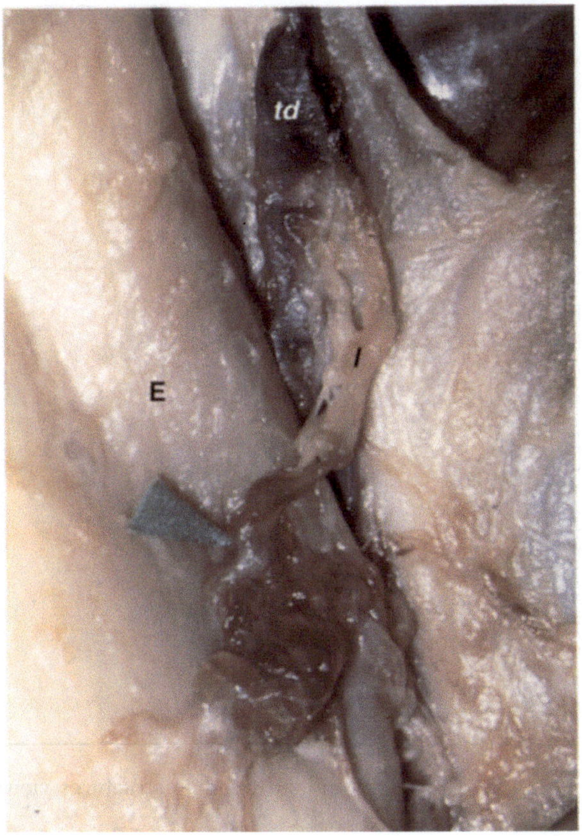

37

Lymph Nodes of the Mediastinum Viewed from the Left

Fig. 38

This illustration shows the relationship of the esophagus to the surrounding structures as viewed from the left (specimen 8). The left lung and bronchus have been shifted to the front. Part of the esophagus is concealed from view by the distal half of the aortic arch. The left intercostal arteries, left bronchial artery, and esophageal artery proper originate from the aorta. The left pulmonary artery (*purple*) has also been reflected. This artery and the aortic arch are connected by the ligamentum arteriosum, around which the left recurrent laryngeal nerve winds toward the back of the aortic arch. Several anterior mediastinal nodes are seen on the surface of the aortic arch. The node which is located adjacent to the ligamentum arteriosum is called the node of the ligamentum arteriosum (node of the duct of Botallo). The lowermost nodes of the left paratracheal chain, the infraaortic arch nodes, located close to the left recurrent laryngeal nerve and deeper than the above-mentioned node of the ligamentum arteriosum, are partially visible here.

Fig. 38. Esophagus and surrounding structures viewed from the left (specimen 8, male)

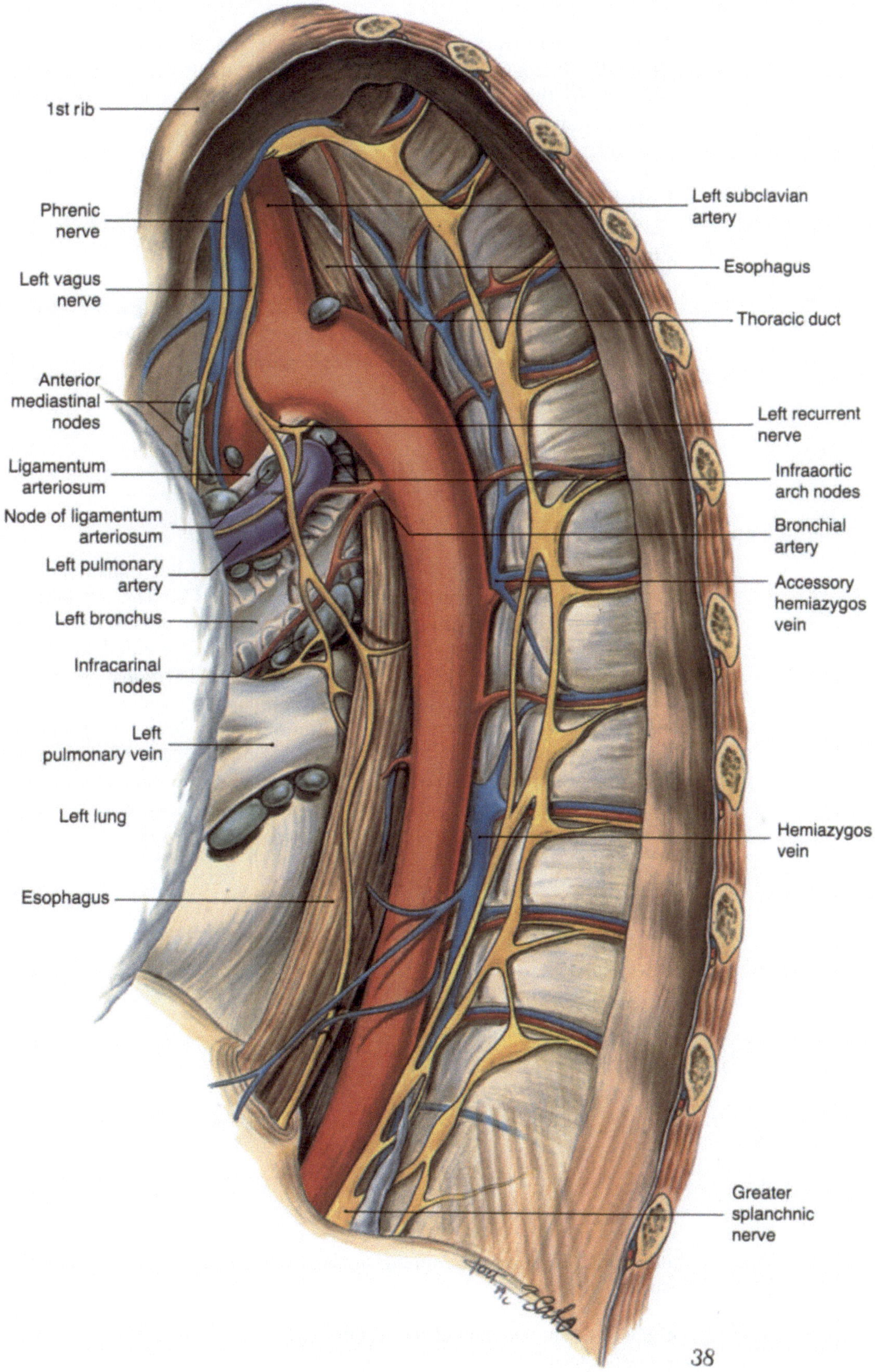
1st rib
Phrenic nerve
Left vagus nerve
Anterior mediastinal nodes
Ligamentum arteriosum
Node of ligamentum arteriosum
Left pulmonary artery
Left bronchus
Infracarinal nodes
Left pulmonary vein
Left lung
Esophagus
Left subclavian artery
Esophagus
Thoracic duct
Left recurrent nerve
Infraaortic arch nodes
Bronchial artery
Accessory hemiazygos vein
Hemiazygos vein
Greater splanchnic nerve

38

Fig. 39

This is a diagrammatic representation of the illustration in Fig. 38. Generally, it is considered difficult to access the left and right paratracheal nodes from this direction. The *striped light green* nodes are the para-esophageal nodes.

Left Superficial Pathway

Fig. 40

In this photograph of specimen 6, the left superficial pathway from the aortic arch to the left venous angle (*upper blue pin*) is shown. The lymph vessel extends parallel to the inferior margin of the left brachiocephalic vein (*lower blue pin*).

Left Cervicothoracic Transitional Region

Fig. 41

This is a photograph of the deep dissection of the left transitional region between the cervical and thoracic areas (specimen 7). The left lobe of the thyroid gland has been pulled to the right. The major vessels have been cut and moved away. The left paratracheal nodes are indicated by the *black arrows*. The left cervical and thoracic paratracheal lymphatics, which are situated between the trachea and the left recurrent laryngeal nerve (*upper and lower right yellow arrows*), communicate via intervening nodes and form the left recurrent nerve chain. A lymph vessel (*series of green arrows*) runs obliquely from the paratracheal nodes to the left supraclavicular nodes. The *green* and *blue pins* indicate the anterior mediastinal lymphatics, and the arch of the azygos vein, respectively.

Fig. 39. Diagram of the illustration in Fig. 38

Fig. 40. Left superficial pathway in the mediastinum to the left venous angle (specimen 6)

Fig. 41. Deep dissection of the lymphatics in the left cervicothoracic transitional area (specimen 7)

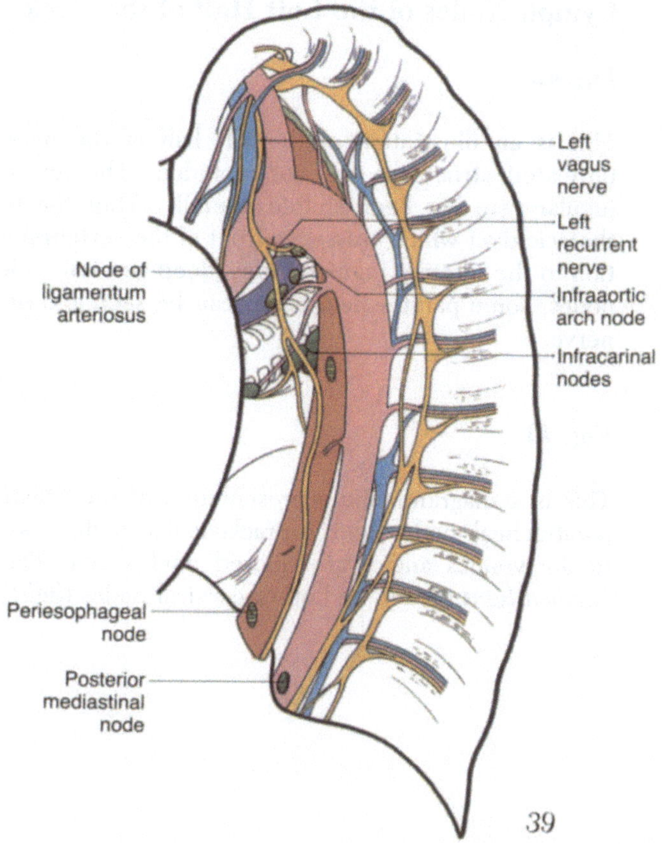
Left vagus nerve
Left recurrent nerve
Infraaortic arch node
Infracarinal nodes
Node of ligamentum arteriosus
Periesophageal node
Posterior mediastinal node

39

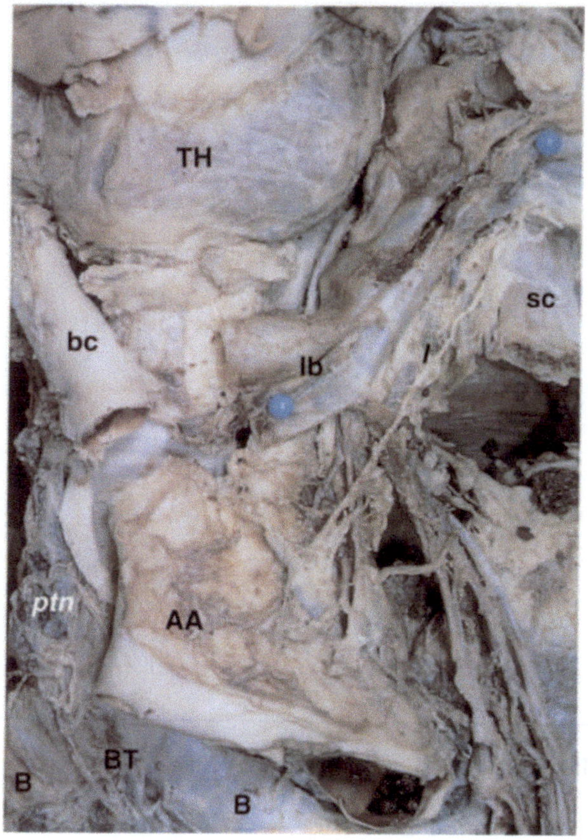
TH
sc
bc
lb
l
ptn
AA
BT
B
B

40

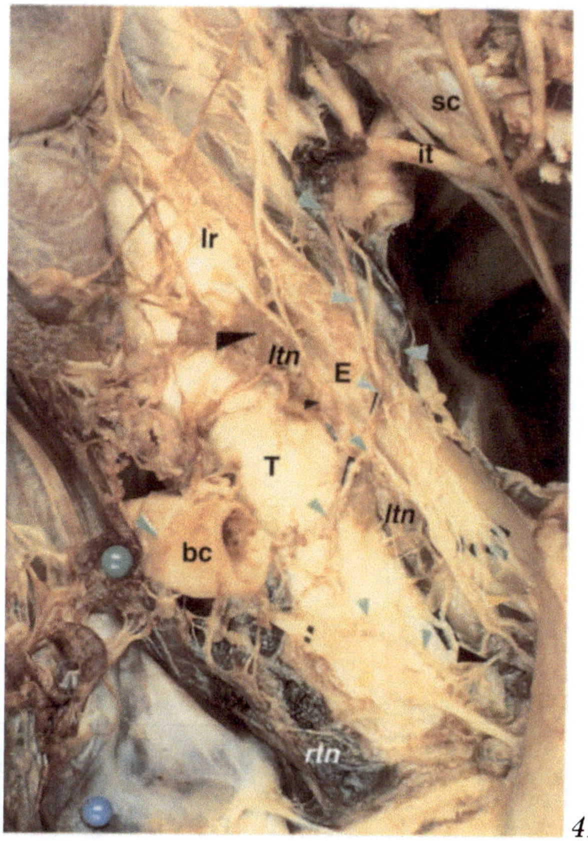
sc
it
lr
ltn
E
T
ltn
bc
rtn

41

Lymph Nodes of the Left Half of the Neck

Fig. 42

This is an illustration of the left half of the neck (specimen 1). Many unrelated structures are not included. The lower part of the internal jugular vein has been shifted laterally. Thus the terminal portion of the thoracic duct which crosses in front of the vertebral vein is visible. In addition to the internal jugular nodes (deep cervical nodes) and supraclavicular nodes, some paratracheal nodes can be seen anterior to the left recurrent nerve.

Fig. 43

This is a diagrammatic representation of the illustration in Fig. 42. The paratracheal nodes and supraclavicular nodes, which are located close to the venous angle, are colored *dark green*. The paraesophageal node (*striped light green*), and other cervical nodes (*light green*) are also shown.

Fig. 42. Lymphatics of the left half of the neck (specimen 1, male)

Fig. 43. Diagram of the illustration in Fig. 41

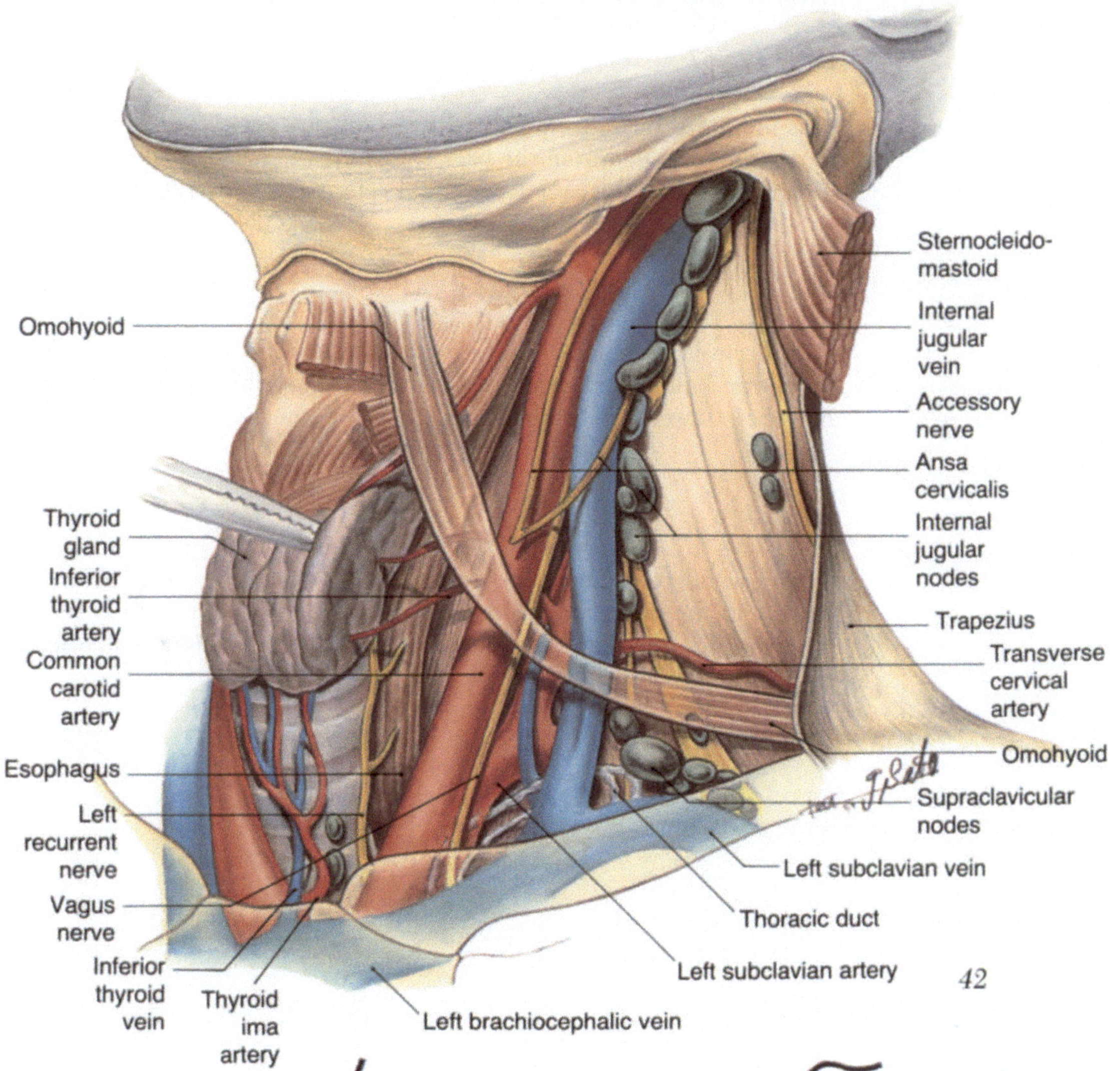
Sternocleido-
mastoid
Omohyoid
Internal
jugular
vein
Accessory
nerve
Ansa
cervicalis
Thyroid
gland
Internal
jugular
nodes
Inferior
thyroid
artery
Trapezius
Common
carotid
artery
Transverse
cervical
artery
Omohyoid
Esophagus
Supraclavicular
nodes
Left
recurrent
nerve
Left subclavian vein
Vagus
nerve
Thoracic duct
Inferior
thyroid
vein
Left subclavian artery
Thyroid
ima
artery
Left brachiocephalic vein

42

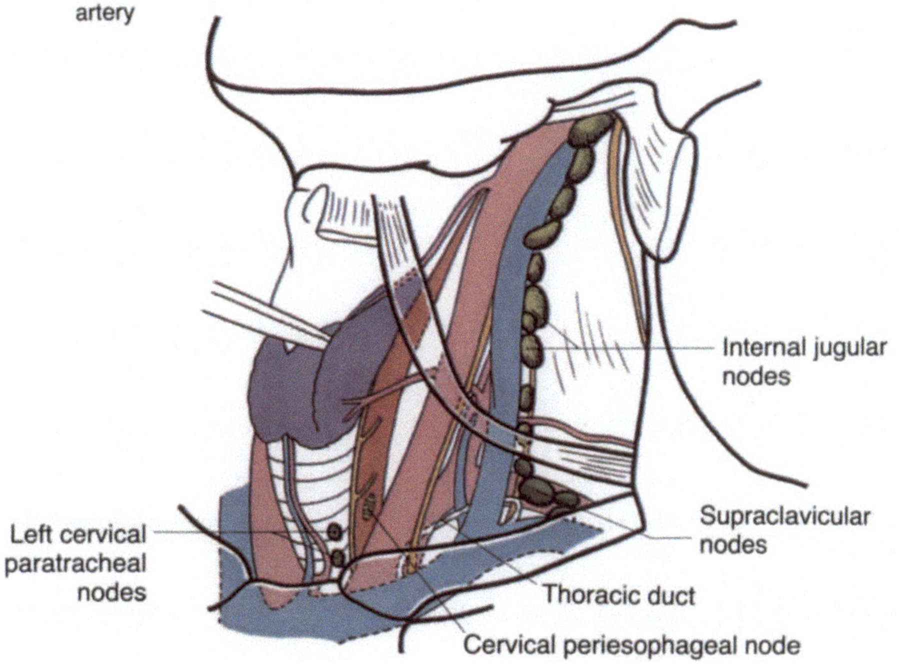
Internal jugular
nodes
Supraclavicular
nodes
Left cervical
paratracheal
nodes
Thoracic duct
Cervical periesophageal node

43

Mediastinal Nodes Viewed from the Back

Figs. 44–67

This unique series of sequential dissection photographs of the mediastinal areas as approached from the back is included to facilitate a comprehensive understanding of this area for surgical reference (specimen 9, male).

Fig. 44

The cervical region from just below the hyoid bone and the area just above the aortic bifurcation have been removed. The ribs and vertebral column have been carefully stripped from the soft tissues, using a sharp scalpel.

Fig. 45

After careful removal of the vertebral column, the cervical prevertebral fascia and the posterior mediastinum, covered by soft tissues between the right and left pleurae, can be seen. The segmental intercostal vessels and nerves are in clear view.

Fig. 44. Dissection of the mediastinal area approached from the back (specimen 9, male)

Fig. 45. Posterior thoracic area, just after removal of the vertebral column (specimen 9)

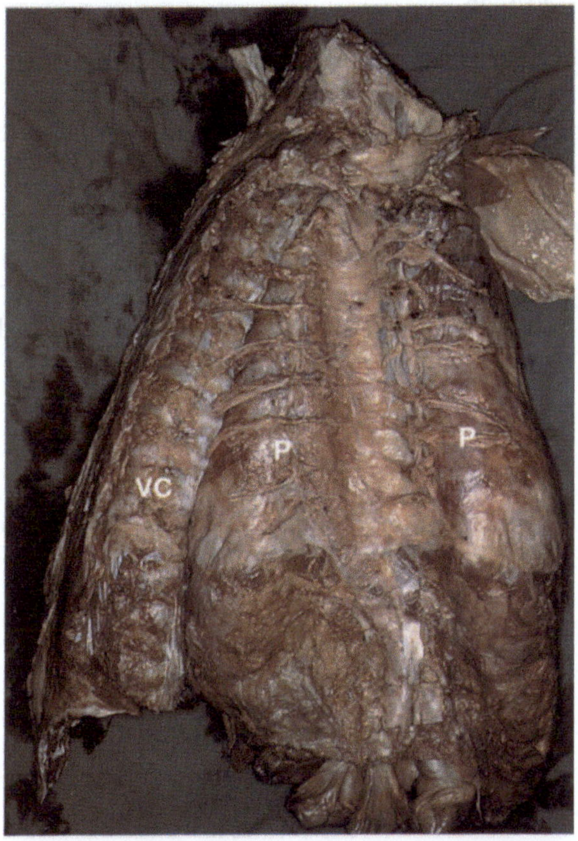

44

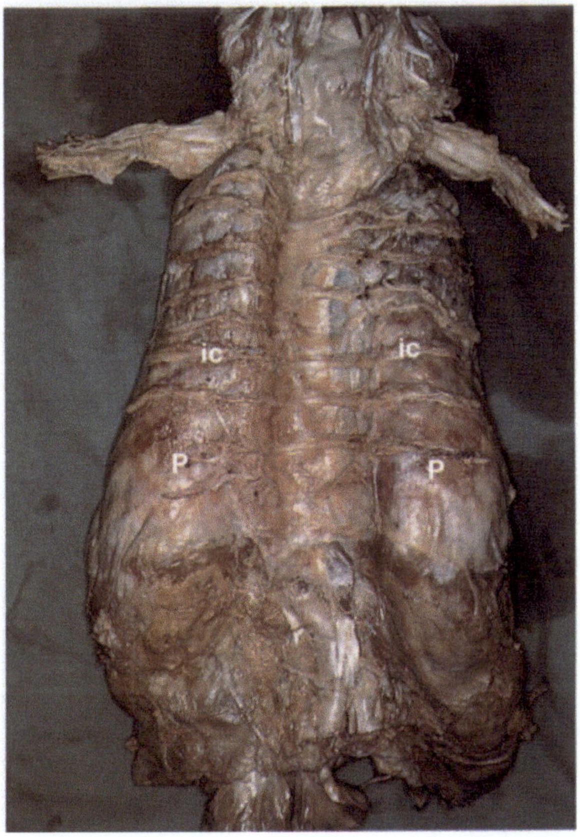

45

Fig. 46

After removal of the intercostal structures, the right and left pleurae were pulled to the outside, allowing clear view of the descending aorta. The azygos and hemiazygos veins run along the aorta.

Fig. 47

The aorta is pulled to the right allowing view of the posterior surface of the esophagus. Note the two esophageal arteries which originate directly from the aorta.

Fig. 48

The azygos vein has been cut and removed. The aorta is shifted to the left so that the thoracic duct can be seen.

Fig. 46. Posterior view of the descending aorta (specimen 9)

Fig. 47. Posterior surface of the esophagus (specimen 9)

Fig. 48. Thoracic duct from behind (specimen 9)

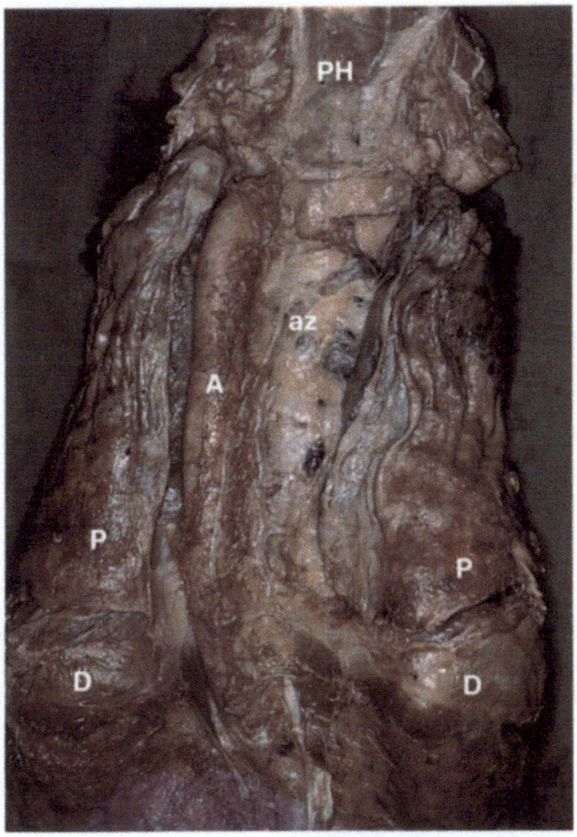

46

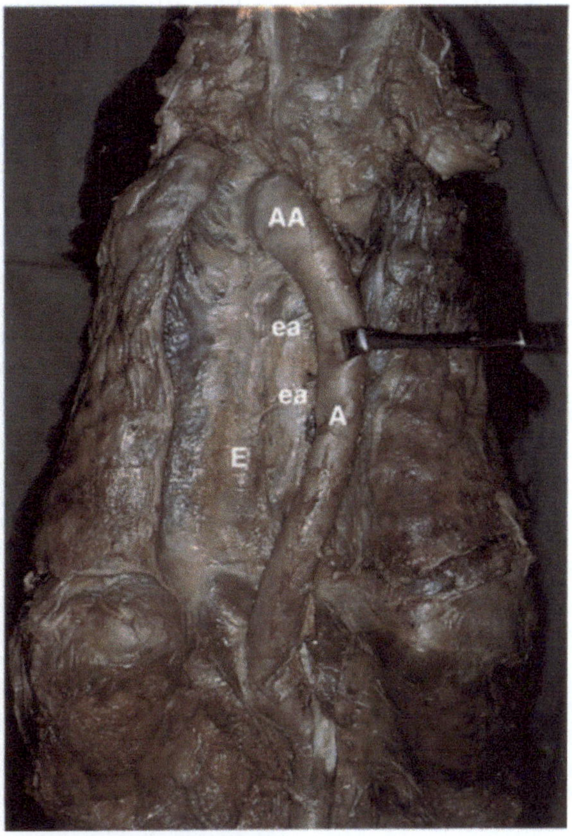

47

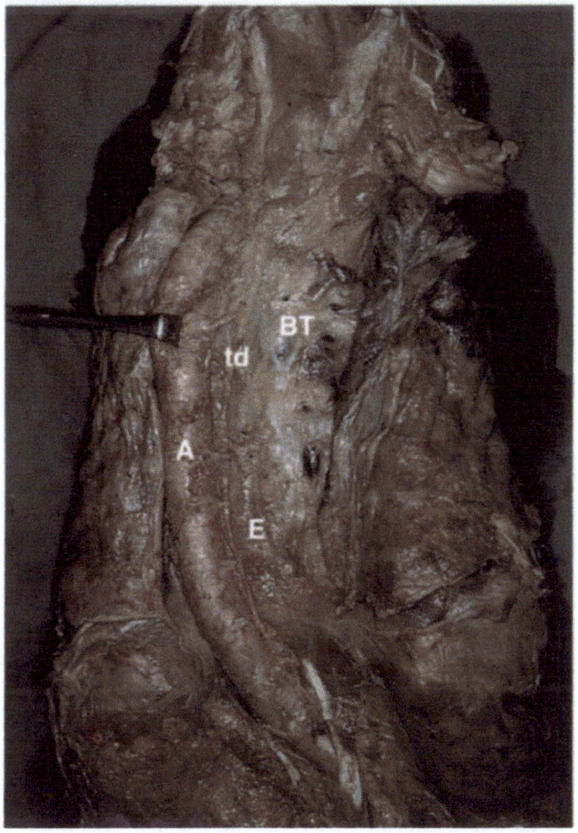

48

Fig. 49

The aorta, which was cut slightly above the diaphragm, is pulled to the left, and the esophagus and the following posterior mediastinal structures can be seen: bronchi, vagus nerve and the posterior vagal trunk, inferior pulmonary veins, pericardium, and the pulmonary ligaments.

Fig. 50

The thoracic duct has been cut above the diaphragm and is reflected upward. With the esophagus slightly shifted to the left, numerous small arteries (*red arrows*) of varying origin can be seen: note those from the right inferior thyroid artery and those from the lower (concave) surface of the aortic arch.

Fig. 51

The esophagus is strongly shifted to the right, allowing view of the trachea, the left recurrent laryngeal nerve, and the infraaortic arch region (area between the aortic arch and the left tracheobronchial angle). Note the lymph vessel which directly connects the esophagus and the thoracic duct at the level of the aortic arch.

Fig. 49. Posterior mediastinal structures (specimen 9)

Fig. 50. Small arteries of the esophagus (specimen 9)

Fig. 51. Posterior surface of the trachea (specimen 9)

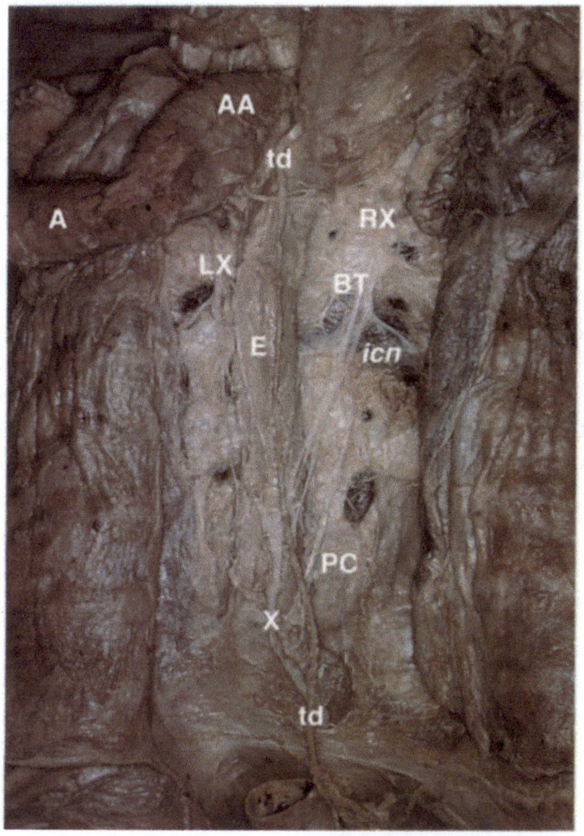

49

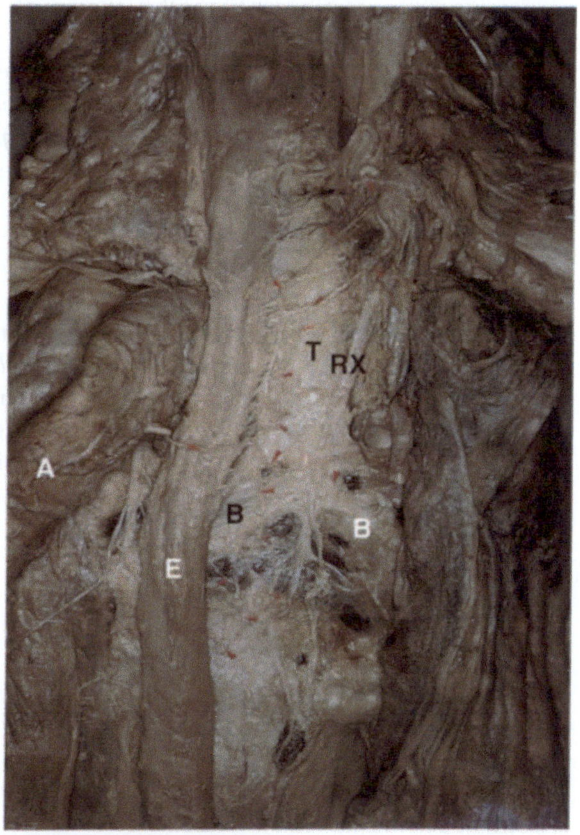

50

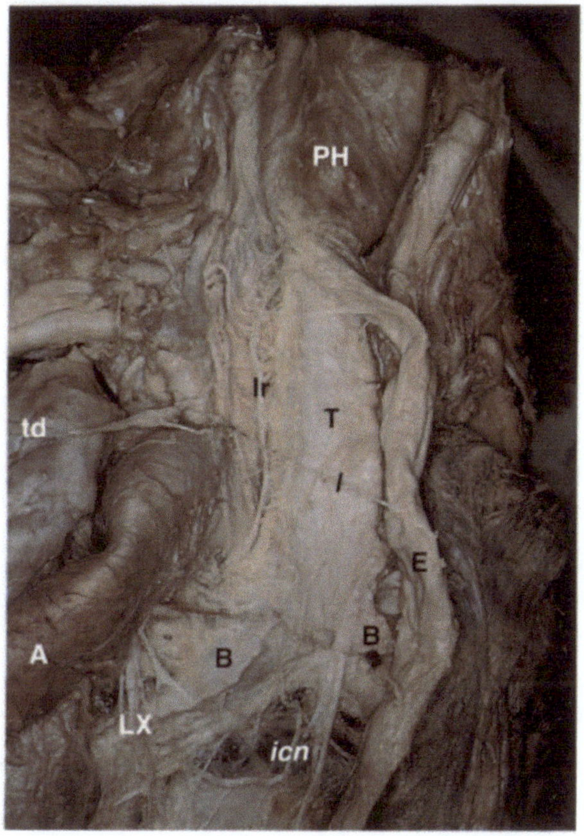

51

Fig. 52

In a closer view, the above-mentioned connecting lymph vessel is highlighted by *inlaid black paper* (*right*). Another communicating lymph vessel (*left inlaid black paper*) from the infraaortic arch area extending to the thoracic duct can be seen.

Fig. 53

Here the above-mentioned right communicating lymph vessel was removed. The *inlaid black papers* show an arterial branch from the highest right intercostal artery of aortic origin which divides into two twigs. The lower twig runs along the posterior surface of the right bronchus and the upper twig communicates with a branch from the right inferior thyroid artery.

Fig. 52. Lymph vessels which drain into the thoracic duct (specimen 9)

Fig. 53. Small arterial twigs on the posterior surface of the trachea (specimen 9)

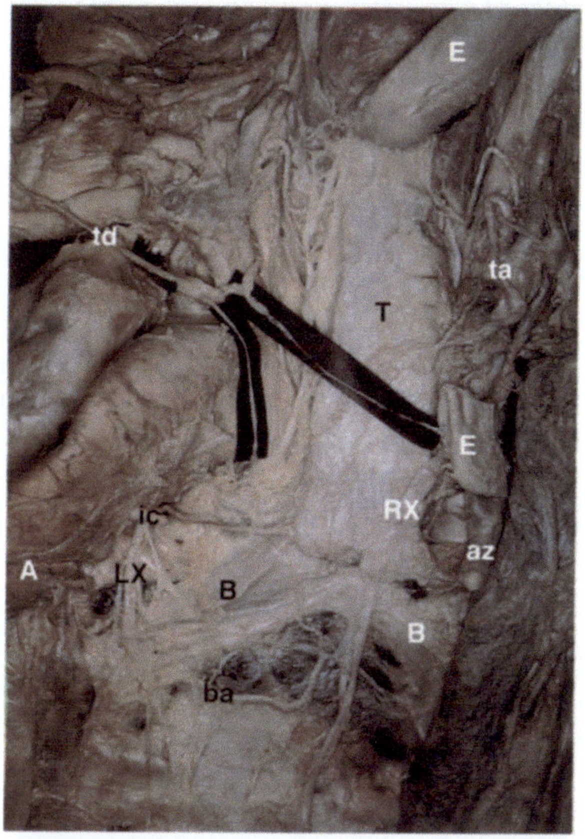

52

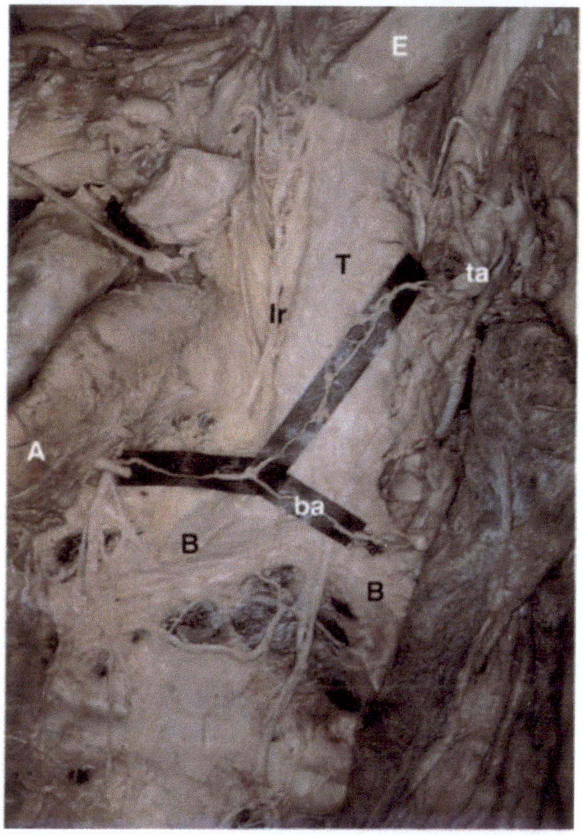

53

Fig. 54

The aorta has been strongly reflected cranially and the above-mentioned arterial twigs have been removed. The right and left vagus nerves descend posterior to their respective bronchus. The communicating branches between these vagi can be seen along the inferior margin of the left bronchus (*inlaid black paper*).

Fig. 55

The above-mentioned communication was cut and the vagus nerves were pulled laterally and pinned (*yellow pins*). Now the infracarinal node group, posterior to the pericardium and anterior to the esophagus, can be seen.

Fig. 54. Communicating branches between the right and left vagus nerves (specimen 9)

Fig. 55. Infracarinal node group (specimen 9)

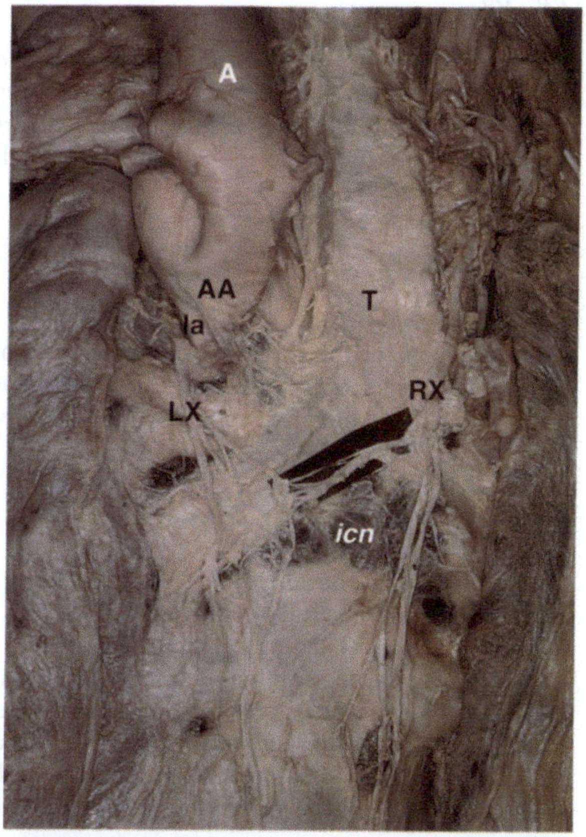

54

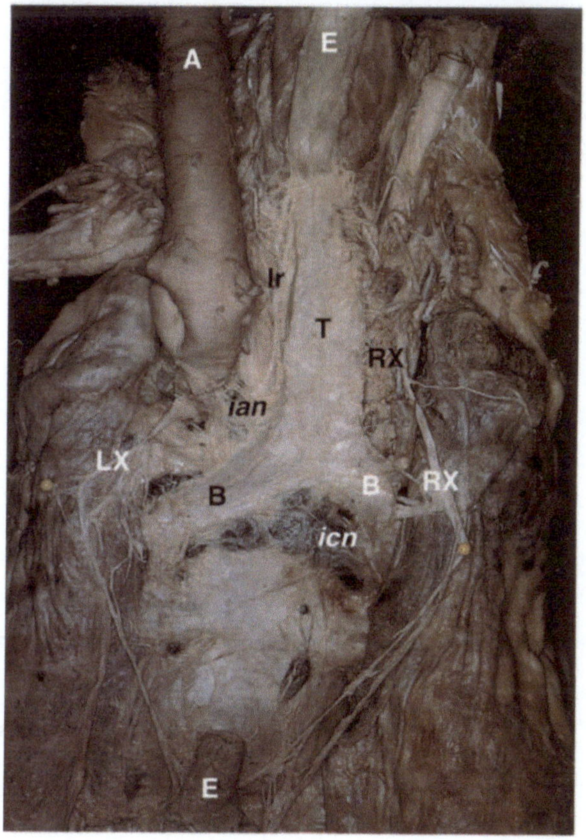

55

Fig. 56

In order to clearly view the infraaortic arch nodes, the aortic arch was cut and fixed (*red pin*) just distal to the ligamentum arteriosum. Note the numerous communicating vessels, nerve network, left recurrent laryngeal nerve, and an artery from the concave surface of the aortic arch to the left bronchus.

Fig. 57

Here the arch of the aorta is pulled more cranially to allow view of the ligamentum arteriosum, the origin of the above-mentioned artery, the vagal trunk, and the left recurrent laryngeal nerve.

Fig. 58

A small section of the aortic arch, which includes the origin of the bronchial artery, was cut, shifted to the left along with the bronchial artery and pinned (*red pin*). The aortic arch was retracted cranially to allow a wider view of the infraaortic arch area, including the ligamentum arteriosum and the initial portion of the left recurrent laryngeal nerve (infraaortic arch nodes) and the numerous networks of nerves and lymphatics. The *black inlaid papers* show communicating lymph vessels, which ascend along the posterior surface of the pericardium and drain into the infracarinal nodes.

Fig. 56. Infraaortic arch area (specimen 9)

Fig. 57. View of part of the ligamentum arteriosum (specimen 9)

Fig. 58. Initial portion of the left recurrent laryngeal nerve and the ascending lymphatics to the infracarinal nodes (specimen 9)

56
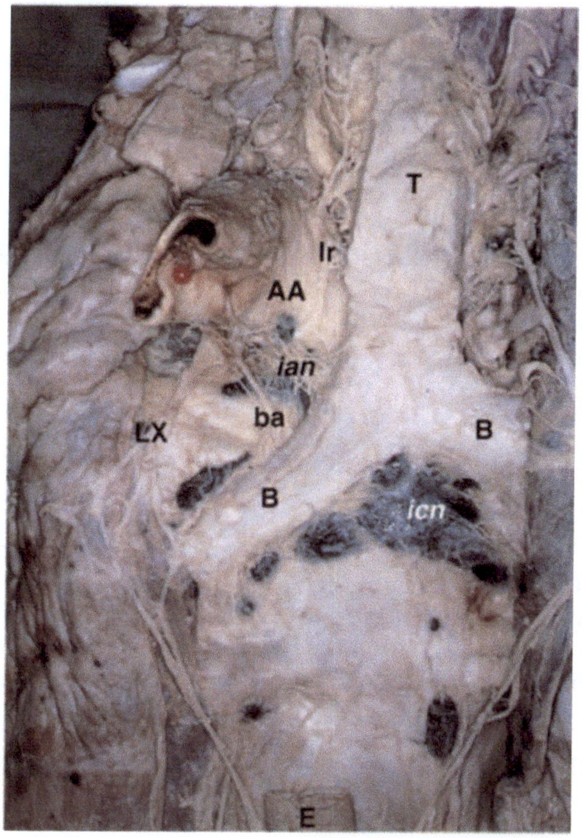
T
lr
AA
ian
LX
ba
B
B
icn
E

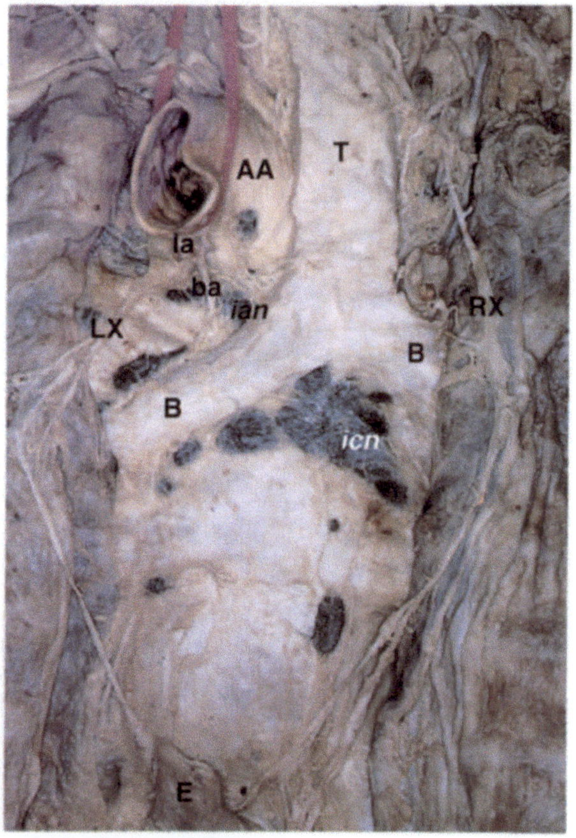
AA
T
la
ba
ian
RX
LX
B
B
icn
E
57

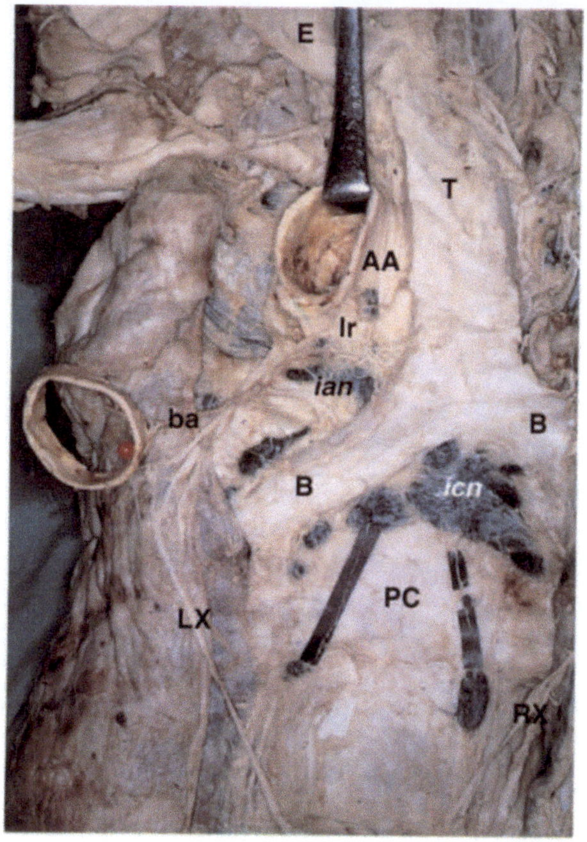
E
T
AA
lr
ian
ba
B
B
icn
PC
LX
RX
58

Fig. 59

Here the left bronchus was carefully peeled away from the soft tissues surrounding it, and a *blue string* was inserted around the bronchus.

Fig. 60

In this photograph, the interbronchial ligament (*blue green arrow*), which lies adjacent to the uppermost part of the infracarinal node group, can be seen. The left bronchus was cut and slightly elevated for subsequent reflection of the trachea to allow view of the anterior surface of the bifurcation area.

Fig. 61

The left bronchus was cut and reflected to the right, and the above-mentioned artery (*ba* in Fig. 58) was cut at its origin from the aortic arch and pinned. This artery extended to pass between the inferior margin of the right bronchus and the infracarinal nodes. The hilar nodes along the upper and lower margins of the left bronchus, as well as one on the anterior surface, can be seen.

Fig. 59. Careful stripping of the left bronchus (specimen 9)

Fig. 60. Interbronchial ligament (specimen 9)

Fig. 61. Nodes on the anterior surface of the left bronchus (specimen 9)

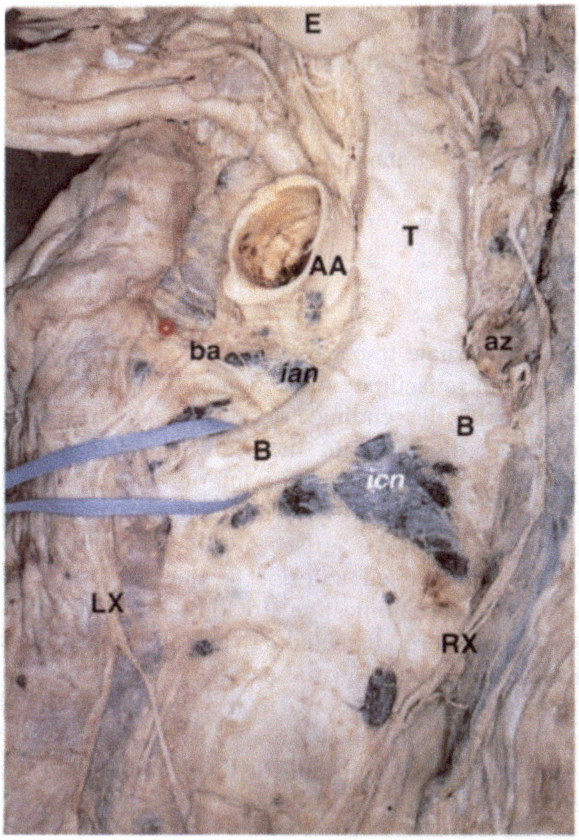

59

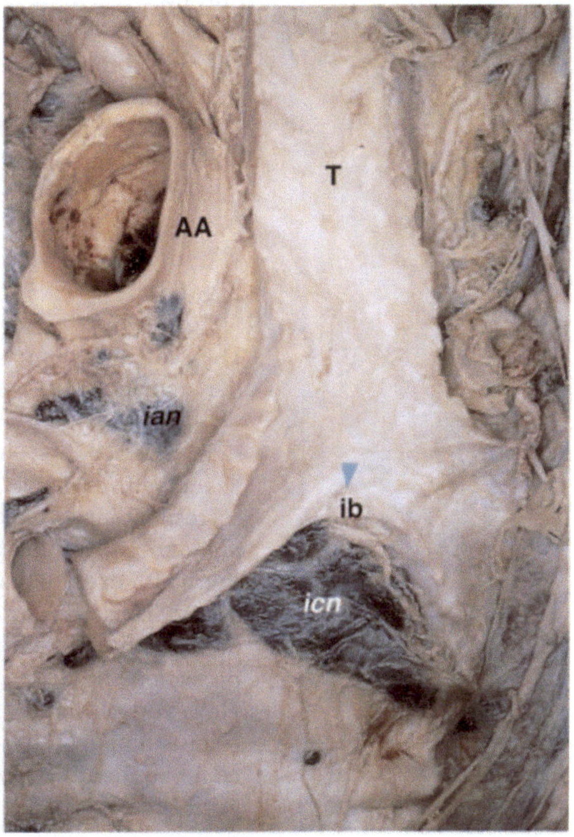

60

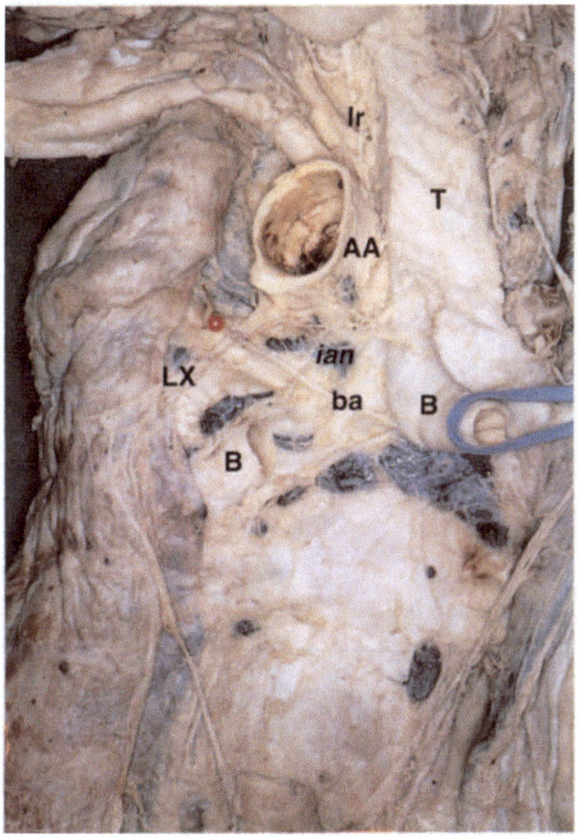

61

Fig. 62

The right bronchus and the lower part of the trachea were carefully peeled away from the surrounding soft tissues and the bronchi were strongly reflected to the right, showing the anterior surface of the trachea. The pretracheal areolar tissues, which contain numerous lymphatics, could subsequently be dissected.

Fig. 63

The right bronchus was cut, the whole trachea was reflected cranially and the areolar tissues were removed to facilitate dissection of the paratracheal vessel and nerves. Note that the descending artery (*red pin*) from the right inferior thyroid artery is divided into the right and left (lesser developed) bronchial arteries, which run along the inferior margin of the corresponding bronchus. The bronchial artery (*center left margin of this photograph*), originating from the concave surface of the aortic arch (ref. Fig. 56), extended obliquely and passed anterior to the left bronchial artery. These two arteries had no communication. In addition to the vagus nerves, the sympathetic nerves were dissected. Even after meticulous dissection, no deep pathway was detected in the paratracheal soft tissues. However, an important intermediate node between the right paratracheal chain (deep pathway) and the brachiocephalic nodes (superficial nodes) could be identified to the right of the brachiocephalic trunk and behind the right brachiocephalic vein.

Fig. 62. Anterior surface of the trachea separated from surrounding soft tissues (specimen 9)

Fig. 63. Arteries and sympathetic nerves between the trachea and the aorta (specimen 9)

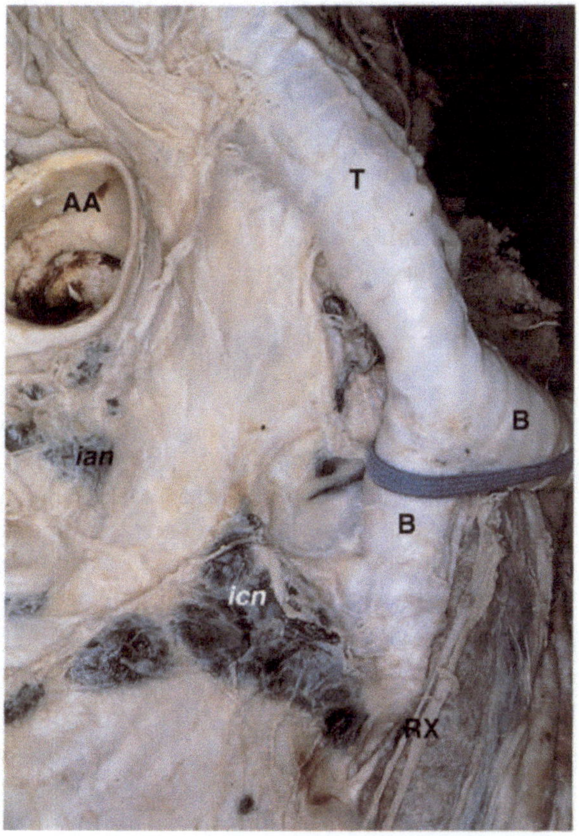

62

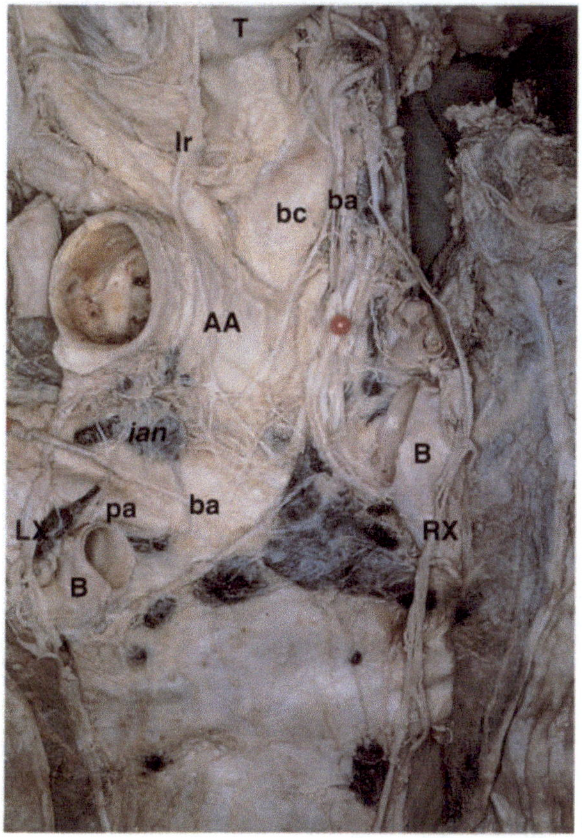

63

Fig. 64

In order to identify any superficial pathways, the whole specimen was completely turned over to allow view from the front. In the *center* of this photograph, an important triangular-shaped space formed by the right and left brachiocephalic veins and the brachiocephalic trunk can be seen. Within this space, numerous brachiocephalic angle nodes were found. These nodes were connected to the right paratracheal chain via the above-mentioned intermediate node (Fig. 63).

Fig. 65

After removal of the soft tissues, numerous superficial communicating vessels from the brachiocephalic angle nodes can be seen. Some vessels ascended superficial to the right brachiocephalic vein to drain into the right supraclavicular nodes. Additionally, some vessels ascending obliquely superficial to the initial portion of the trachea received lymphatics from the thyroid gland and reached the left venous angle.

Fig. 66

After reflection of the superior vena cava to the right and upwards, and the aortic arch to the left, the above-mentioned brachiocephalic angle node connections can be seen from the front at a slightly oblique angle. The complete pathway from the right paratracheal nodes to the right venous angle can now be clearly viewed; this is the connection between the right deep and superficial pathways via the brachiocephalic angle nodes.

Fig. 64. Brachiocephalic angle area from the front (specimen 9)

Fig. 65. Brachiocephalic angle nodes and numerous communicating vessels to both the right and left venous angle (specimen 9)

Fig. 66. Brachiocephalic angle nodes serve as the connection between the right deep and superficial pathways (specimen 9)

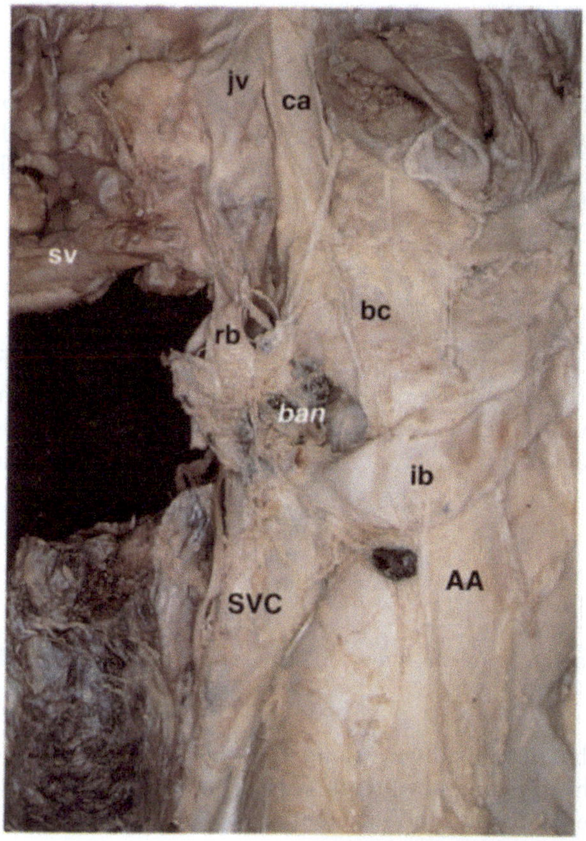
jv
ca
sv
rb
bc
ban
ib
SVC
AA

64

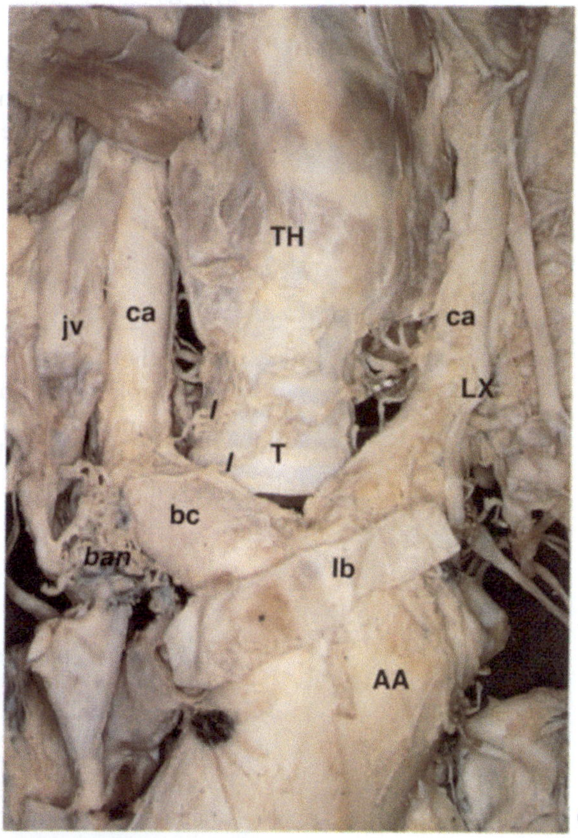
TH
jv
ca
ca
LX
l
l
T
bc
ban
lb
AA

65

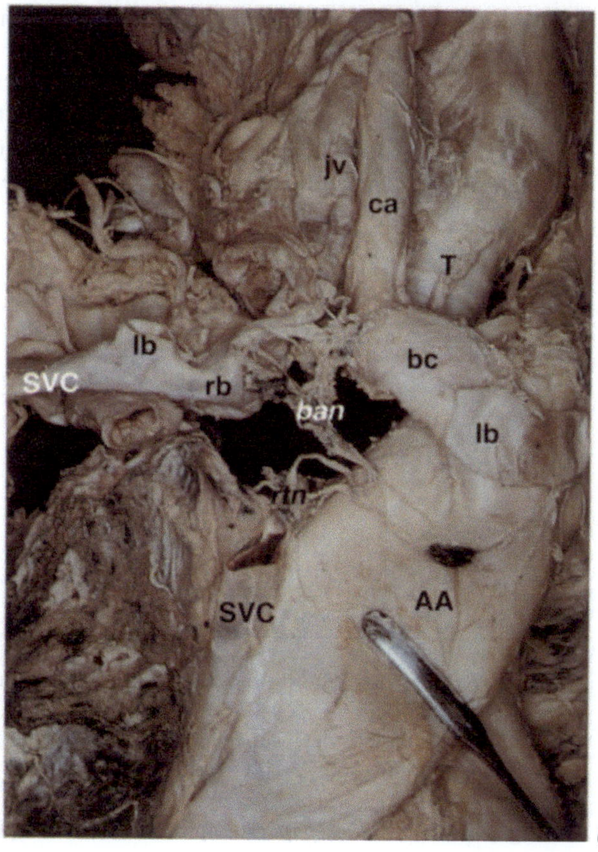
jv
ca
T
lb
SVC
rb
bc
ban
lb
rtn
SVC
AA

66

Fig. 67

Here, viewed from the left, the origin of the left recurrent laryngeal nerve has been elevated in order to allow clear view of the areas both posterior and anterior to the ligamentum arteriosum. After removal of the soft tissues, communicating vessels can be seen between the infraaortic arch nodes and the node of the ligamentum arteriosum.

Fig. 68

Here, the area of Fig. 66 is viewed from the front in order to identify the complete left superficial pathway. From the node of the ligamentum arteriosum, the communicating vessel ascended superficial to the aortic arch and the left subclavian artery via intercalate nodes, and reached the left venous angle.

Fig. 69

Inlaid black papers facilitate clear view of the above-mentioned left superficial pathway.

In this specimen, the deep pathway along the trachea was replaced by the superficial pathway along the great vessels, at the level of the brachiocephalic triangle (*right*) and the ligamentum arteriosum.

In this paper, the deep and superficial pathways of both sides of the tracheo-esophageal region, as well as the most critical nodes of each pathway, were identified. The topographical relationship to the surrounding structures was discussed for the following critical nodes: right recurrent nerve nodes, infraaortic arch nodes, infracarinal nodes, brachiocephalic angle nodes, and nodes of the ligamentum arteriosum.

Fig. 67. Area surrounding the ligamentum arteriosum from the left (specimen 9)

Fig. 68. Complete left superficial pathway from the nodes of the ligamentum arteriosum, viewed from the front (specimen 9)

Fig. 69. *Inlaid black papers* emphasize the left superficial pathway (specimen 9)

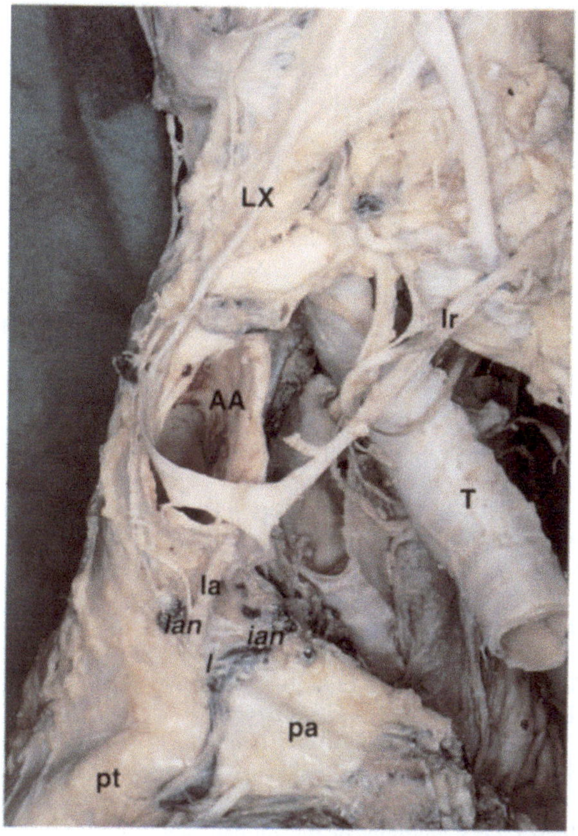

67

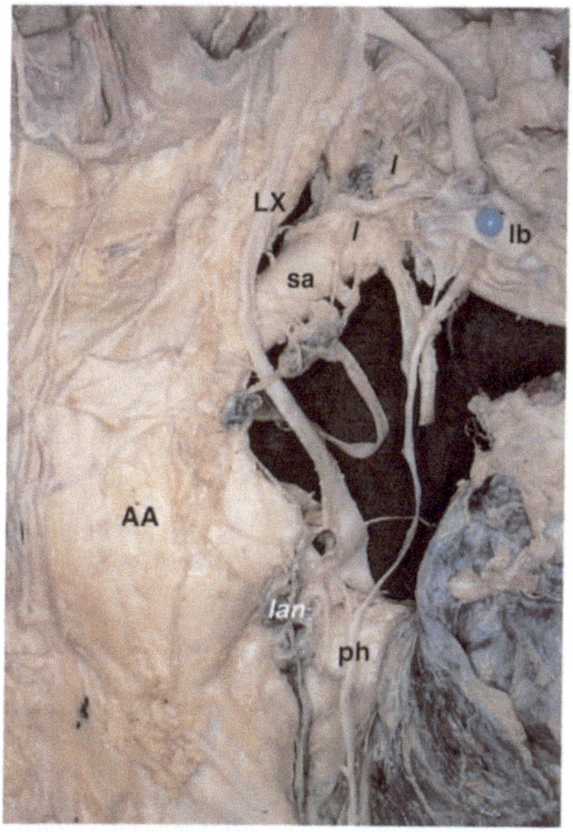

68

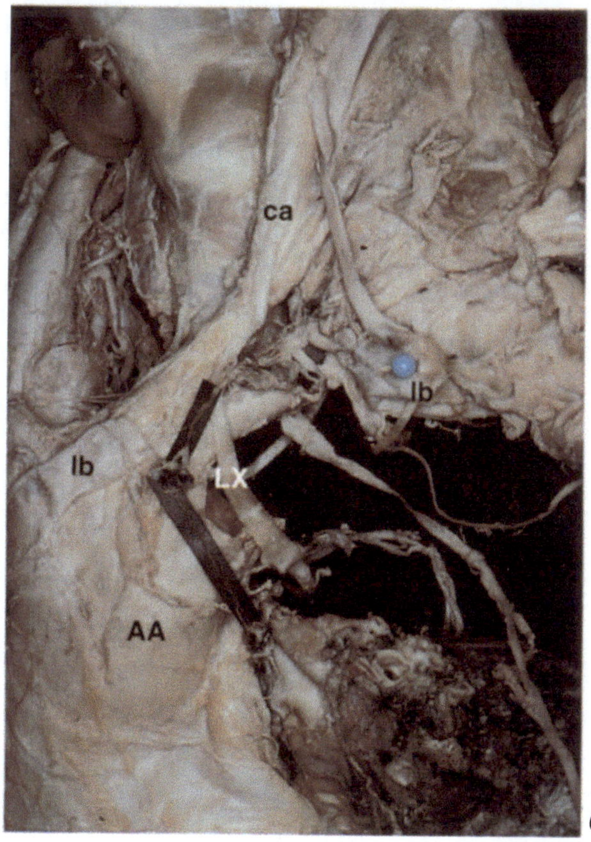

69

Acknowledgment

The authors wish to express their sincere thanks to Professor Emeritus K. Inokuchi for his encouragement and support.

References

1. Poirier P, Cunéo B (1902) System lymphatique. In: Poirier P, Charpy A (eds) Traité d'anatomie humaine, 2nd edn, vol 2. Masson, Paris, pp 1229–1236, 1264–1277
2. Bartels P (1909) Das Lymphgefässsystem. Gustav Fischer, Jena, pp 110–118, 172–179
3. Rouvière H (1932) Anatomie des lymphatiques de l'homme. Masson, Paris, pp 35–59, 168–189
4. Rouvière H, Tobias MJ (1938) Anatomy of the human lymphatic system. Edwards Brothers, Ann Arbor, pp 16–28, 91–101
5. Kutsuna M (1968) Anatomy of the lymphatic system of the Japanese (in Japanese). Kanehara Shuppan, Tokyo, pp 73–77, 116–122
6. Haagensen CD, Feind CR, Herter FP, Slanetz Jr CA, Weinberg JA (1972) The lymphatics in cancer. WB Saunders, Philadelphia, pp 73–86, 251–262
7. International Anatomical Nomenclature Commitee (1983) Nomina anatomica, 5th edn. Williams and Wilkins, Baltimore, pp A61–62
8. Japanese Society for Esophageal Disease (1984) Guidelines for clinical and pathological studies on carcinoma of the esophagus, 6th edn (in Japanese). Kanehara, Tokyo, pp 14–17
9. Clemente CD (ed) (1985) Gray's anatomy of the human body, 30th American edn. Lea and Febiger, Philadelphia, pp 884–887, 894–896
10. Akiyama H (1990) Surgery for cancer of the esophagus. Williams and Wilkins, Baltimore, pp 21–23
11. Kinoshita I, Ohashi I, Nakagawa K, Kajitani T, Kaneda K, Tsuya A (1976) Lymph node metastasis in esophageal cancer; with special reference to upper mediastinum and measures for its treatment (in Japanese). Jpn J Gastrointest Surg 9: 424–430
12. Nishihira T, Ohmori N, Hirayama K, Hyouba R, Sato A (1986) Significance of lymph node dissection in the cancer of the thoracic esophagus (in Japanese). Geka Shinryo (Tokyo) 28: 549–554
13. Engel S (1926) Die Topographie der bronchialen Lymphknoten und ihre präparatorische Darstellung. Beitr Klin Tuberculose 64: 468–481
14. Sato T (1990) Clinical anatomy for esophagus cancer operation (in Japanese). Geka Shinryo (Tokyo) 32: 902–911

Illustrations of Surgery for Carcinoma in the Thoracic Esophagus

Teruo Kakegawa, Hiromasa Fujita, and Hideaki Yamana[1]

Introduction

Since the establishment of the Japanese Society for Esophageal Diseases in 1965 and the publication of the *Guidelines for Clinical* and *Pathologic Studies on Carcinoma of the Esophagus* [1] in 1969, lymph node dissection during esophagectomy has gradually become accepted in Japan, though not so in the rest of the world, perhaps because of the complexity of the procedure. Since cervico-thoraco-abdominal lymph node dissection for carcinoma in the thoracic esophagus was first reported by Sannohe [2] in 1981, this procedure has been adopted as the standard approach. However, there remain many problems in nomenclature and topographical definition of the cervical and mediastinal lymph nodes in the *Guidelines*. Here, we describe in detail our standard procedure for cervico-thoraco-abdominal lymph node dissection, using the nomenclature and definitions set out by the International Society for Diseases of the Esophagus (ISDE) Research Committee [3].

Lymph Nodes for Dissection in Thoracic Esophageal Carcinoma

Formats for describing surgical records involving the nomenclature and definition of regional lymph nodes have been set out by the ISDE Research Committee (Fig. 1 and Table 1).

We studied the number of dissected lymph nodes removed in each region by thoraco-abdominal and cervico-thoraco-abdominal dissection at the First Department of Surgery, Kurume University Hospital (Fig. 2). The number of lymph nodes in the upper mediastinal region was found to be more for cervico-thoraco-abdominal dissection than for thoraco-abdominal dissection. On the other hand, the lymphatic recurrence study showed that cervical and upper mediastinal lymph node recurrence was frequent following thoraco-abdominal dissection (Fig. 3). Such findings led us to adopt a policy of performing cervico-thoraco-abdominal dissection for

[1] First Department of Surgery, Kurume University School of Medicine, Kurume, 830 Japan

Table 1. ISDE Research Committee compilation of lymph node metastasis determined by macroscopic and pathology examinations. (Modified from registration form of esophageal carcinoma of the ISDE TNM Research Committee, 1988).

Lymph nodes	*(Macroscopic)*		*(Microscopic)*
Cervical			
Cervical paraesophageal	yes[1]	no	unknown[2]
Cervical paratracheal	yes	no	unknown
Internal jugular (right)	yes	no	unknown
Internal jugular (left)	yes	no	unknown
Supraclavicular (right)	yes	no	unknown
Supraclavicular (left)	yes	no	unknown
Thoracic			
Periesophageal	yes	no	unknown
Right recurrent nerve nodes	yes	no	unknown
Right paratracheal	yes	no	unknown
Infracarinal	yes	no	unknown
Left paratracheal	yes	no	unknown
Infraaortic arch	yes	no	unknown
Lower posterior mediastinal[3]	yes	no	unknown
Abdominal			
Right paracardiac	yes	no	unknown
Left paracardiac	yes	no	unknown
Lesser curvature	yes	no	unknown
Greater curvature	yes	no	unknown
Left gastric	yes	no	unknown
Common hepatic	yes	no	unknown
Splenic	yes	no	unknown
Celiac	yes	no	unknown
Abdominal paraaortic	yes	no	unknown

[1] The number of positive nodes is the numerator; the number of lymph nodes dissected is the denominator

[2] "Unknown" indicates no pathology examination

[3] Located below infracarinal nodes, excluding periesophageal nodes

thoracic esophageal carcinoma. The results disclosed that the frequency of bilateral cervical and mediastinal lymph node metastasis, especially of the recurrent pharyngeal nerve lymphatic chain, was high after cervico-thoraco-abdominal dissection (Fig. 4), indicating that this approach is valid for radical esophagectomy.

Fig. 1. Nomenclature and topographical definition of lymph nodes for surgical dissection established by the ISDE Research Committee in 1989 (from [3])

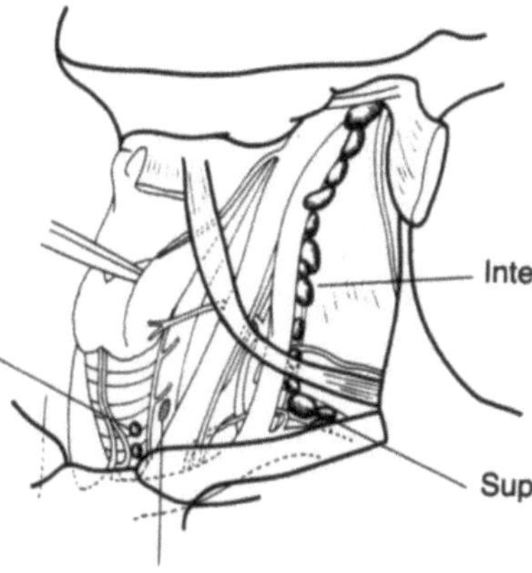
(Cervical approach)
Internal jugular nodes
Cervical paratracheal
Supraclavicular nodes
Cervical paraesophageal

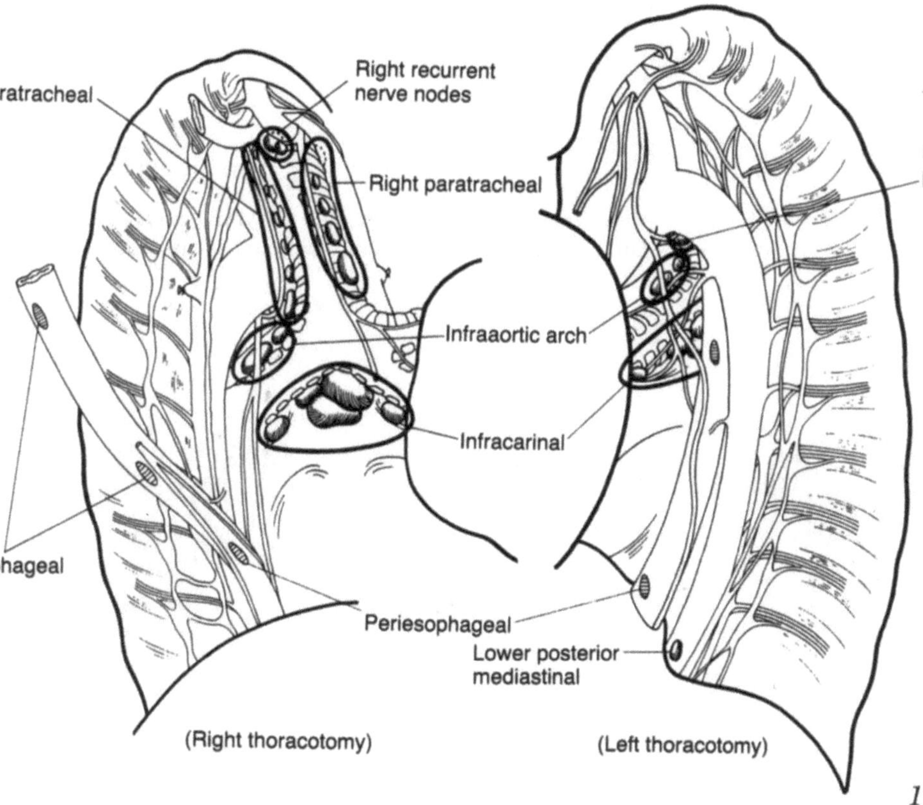
Left paratracheal
Right recurrent nerve nodes
Right paratracheal
Left paratracheal
Infraaortic arch
Infracarinal
Periesophageal
Periesophageal
Lower posterior mediastinal
(Right thoracotomy)
(Left thoracotomy)

1

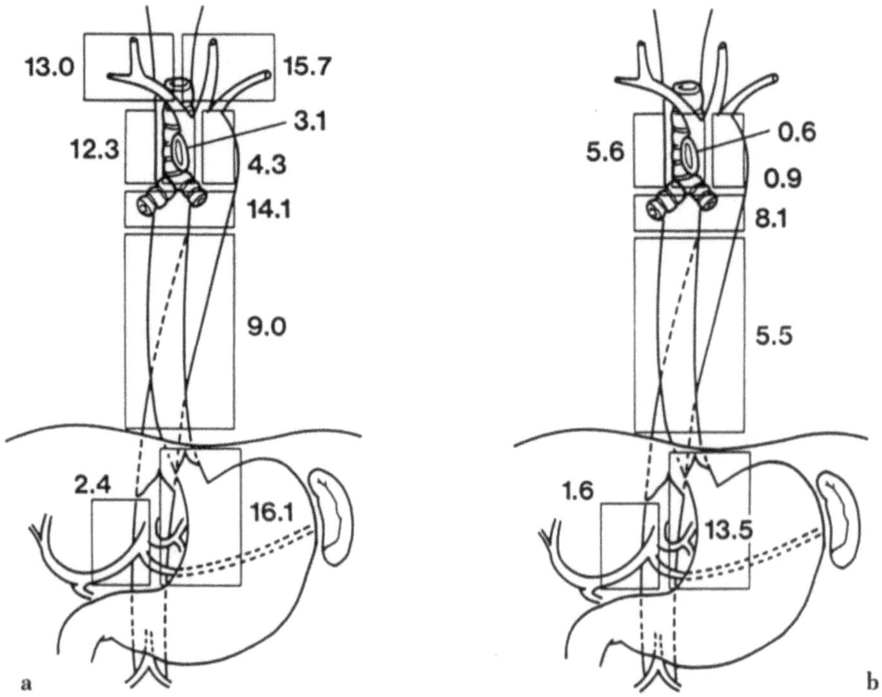

Fig. 2. Comparison between the number of lymph nodes removed per patient **a** cervico-thoraco-abdominal by **b** thoraco-abdominal dissection

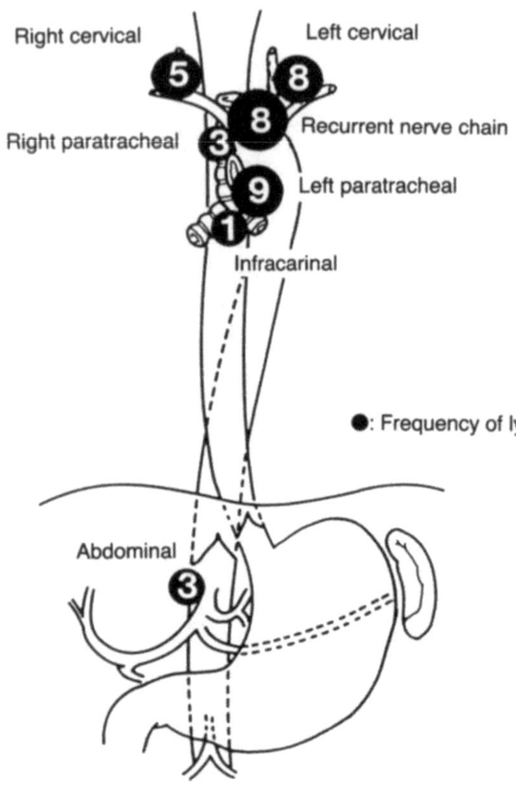

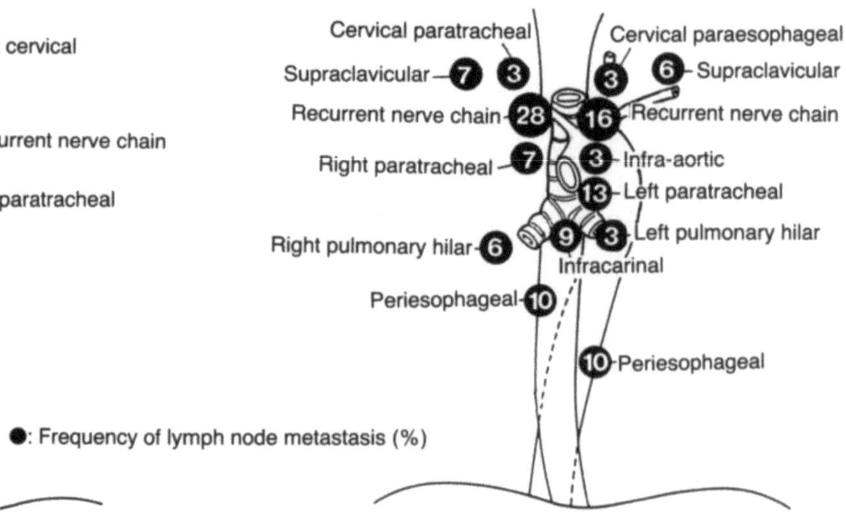

Fig. 4. Incidence rate of lymph node metastasis in cervico-thoraco-abdominal dissection

◁ **Fig. 3.** Frequency of cancer recurrence in regional lymph nodes after thoracoabdominal dissection

Surgical Procedure

Approach for Thoracotomy

The patient lies in the left lateral position. A right posterolateral incision is made between the 5th and 6th ribs. Following transection of these ribs in the posterior portion, the chest is opened through the 5th intercostal space.

Dissection of the Right Recurrent Nerve Nodes (Highest Mediastinal Nodes)

An incision is made into the mediastinal pleura along the superior vena cava, and the right brachiocephalic vein and the right subclavian artery is exposed. The right vagus nerve is then taped. The right recurrent nerve is found when the fatty tissue inferior to the subclavian artery and posterior to the vagus nerve is isolated from the vessel and the nerve. The right recurrent nerve nodes (highest mediastinal nodes) situated posterior to the nerve are removed together with the fatty tissue, being pulled out with a clamp (Fig. 5). These nodes connect with the cervical paratracheal and paraesophageal nodes (the recurrent nerve lymphatic chain) superior to the right subclavian artery, which nodes can be removed later in the cervical procedure. Branches of the right inferior thyroidal artery are then ligated with hemoclips.

Dissection of the Right Paratracheal Nodes

The mediastinal pleura lateral to the superior vena cava is incised up to the azygos vein. After the right vagus nerve is retracted backwards with a tape, the right paratracheal nodes situated between the superior vena cava and the trachea are dissected (Fig. 6). They are removed together with the fatty tissue on the right side of the brachiocephalic artery. During this procedure, the paratracheal sheath should be preserved and the right recurrent nerve should be safeguarded by posterior traction of the vagus nerve with tape. This dissection is performed in a downward direction to the azygos vein.

The azygos vein is now ligated and divided near the superior vena cava. The right bronchial artery is present immediately posterior to the azygos vein. The bronchial artery is taped and preserved to prevent postoperative bronchial ischemia. One or more lymph nodes along the bronchial artery are removed if they are present. The azygos vein is again ligated and resected near the thoracic vertebra. The right tracheobronchial nodes, situated between the right tracheobronchial angle and the superior vena cava and connected with the lower portion of the right paratracheal nodes, are removed (Fig. 7). The tracheal or bronchial branches of the right vagus nerve should be preserved as much as possible (Fig. 8). An ultrasonic aspirator is very convenient during this maneuver.

The merit of preservation of the bronchial artery and the tracheobronchial branches of the right vagus nerve was studied experimentally in dogs. Tracheal blood flow was satisfactorily preserved by this procedure compared to extended ablation of the cervico-thoracic trachea (Fig. 9). In a macroscopic study on tracheal ischemia, experimental models without preservation of the bronchial artery showed moderate to marked tracheal ulceration (Fig. 10). On the other hand, preservation of the tracheobronchial branches of the vagus nerve increased cardiorespiratory tolerance to over-hydration following radical esophagectomy (Fig. 11).

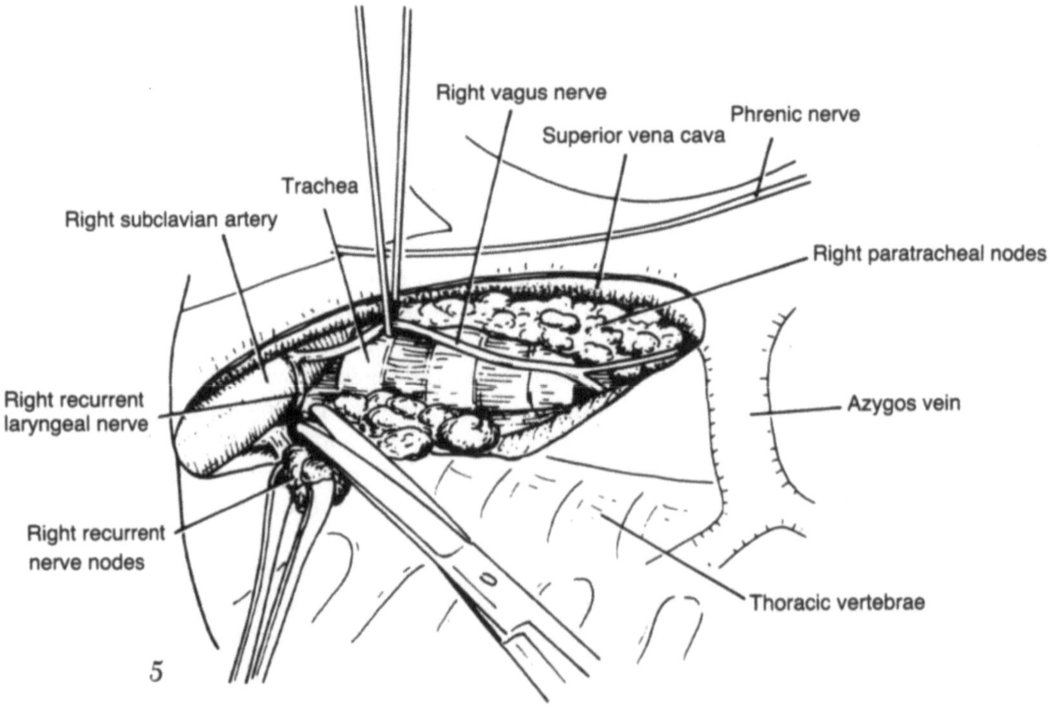

5

Fig. 5. Dissection of the right recurrent nerve nodes (from [4])

Fig. 6. Dissection of the right paratracheal nodes (from [4])

Fig. 7. Dissection of the lower portion of the right paratracheal nodes (from [4])

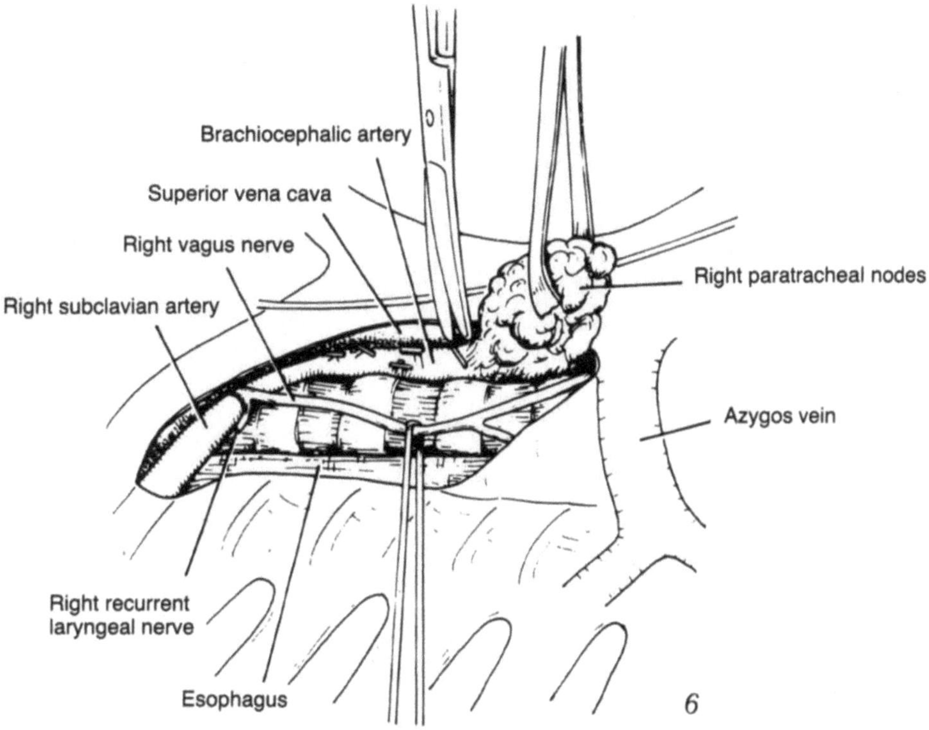
Brachiocephalic artery
Superior vena cava
Right vagus nerve
Right subclavian artery
Right paratracheal nodes
Azygos vein
Right recurrent laryngeal nerve
Esophagus

6

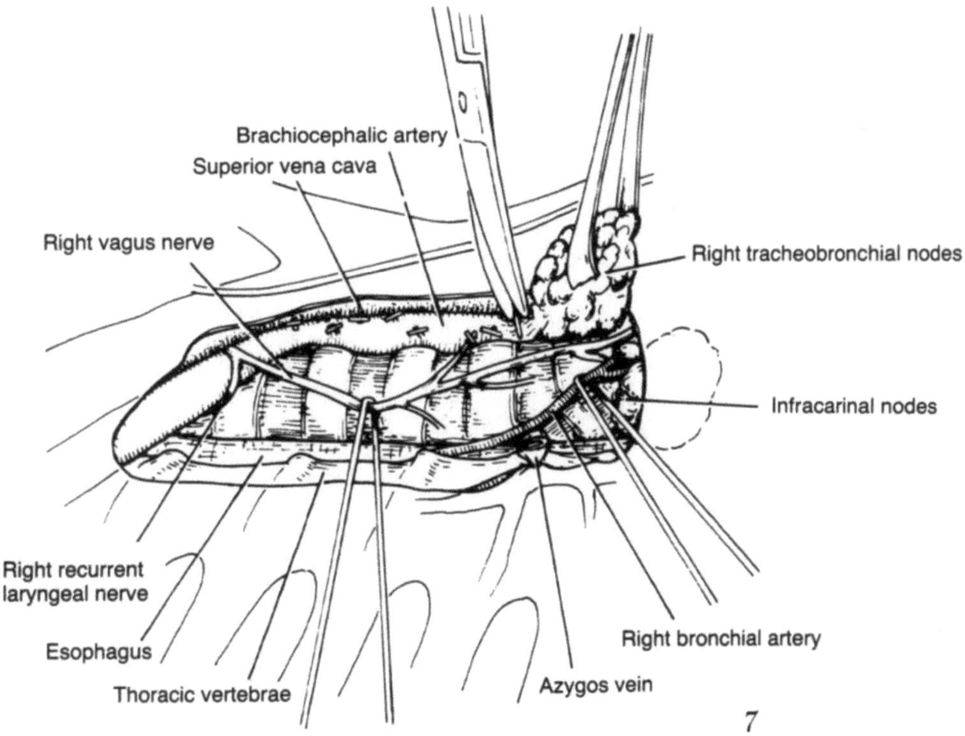
Brachiocephalic artery
Superior vena cava
Right vagus nerve
Right tracheobronchial nodes
Infracarinal nodes
Right recurrent laryngeal nerve
Esophagus
Thoracic vertebrae
Right bronchial artery
Azygos vein

7

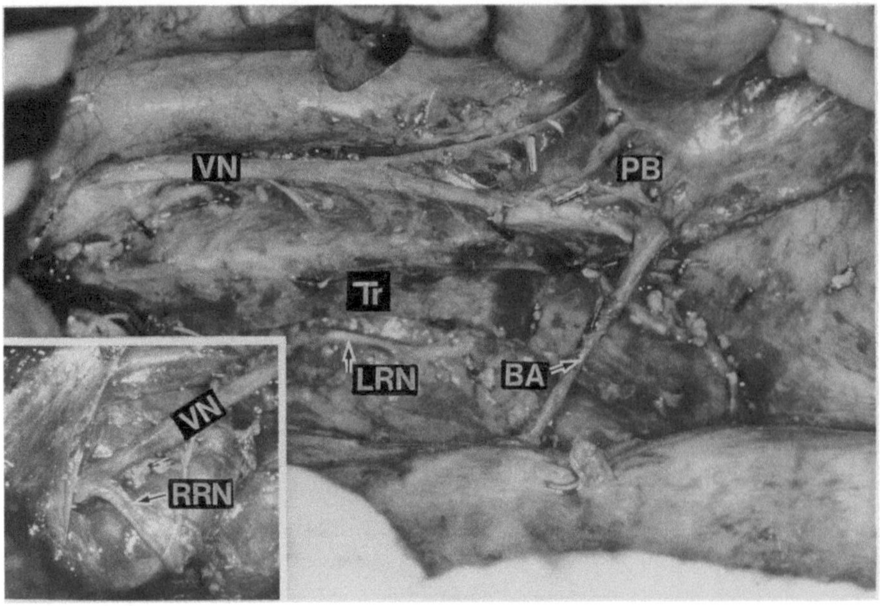

Fig. 8. Intraoperative findings showing preservation of the right bronchial artery and tracheobronchial branches of the vagus nerve. *VN*, vagus nerve; *Tr*, trachea; *PB*, pulmonary branches of the vagus nerve; *BA*, right bronchial artery; *RRN*, right recurrent pharyngeal nerve; *LRN*, left recurrent pharyngeal nerve

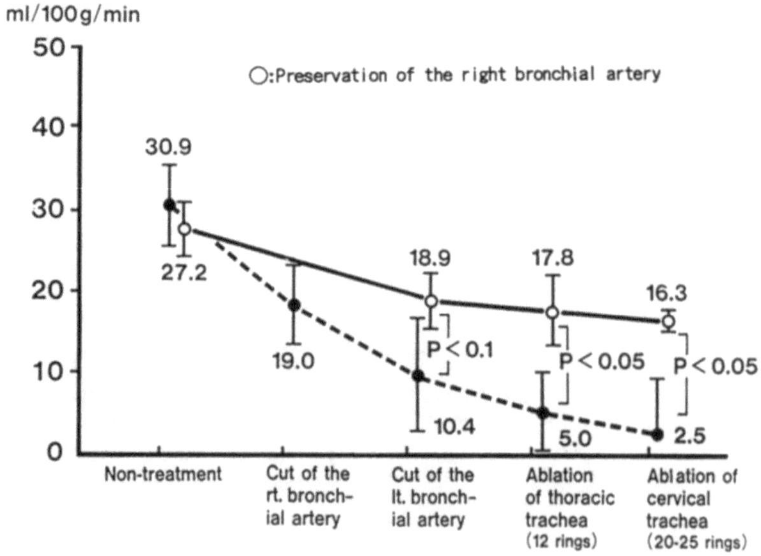

Fig. 9. Changes in tracheal tissue blood flow before and after tracheal devascularization in dogs

Fig. 10. Tracheal ulcerations in the dog (*arrows*) after radical esophagectomy without preservation of the right bronchial artery

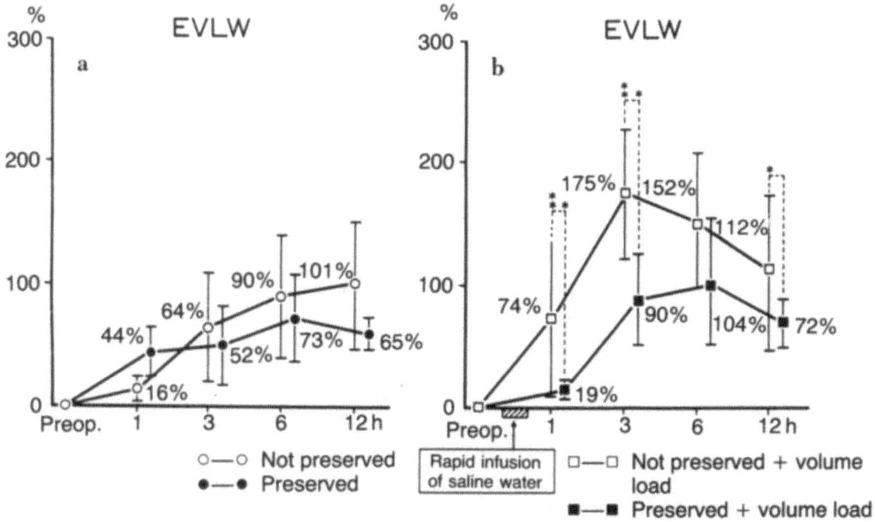

Fig. 11a,b. Changes in extravascular lung water after radical esophagectomy in dogs, with and without preservation of the pulmonary branches of the vagus nerve, **a** without volume load, **b** with volume load. $^{*}P < 0.05$ $^{**}P < 0.01$

Dissection of the Lower Posterior Mediastinal Nodes

The right inferior pulmonary ligament is divided and the posterior mediastinal nodes in the ligament are removed. Following upward retraction of the lower lobe of the right lung, the mediastinal pleura is incised along the posterior margin of the pericardium and the anterior margin of the decending aorta. After isolation from the pericardium, as much as is possible, the esophagus is, taped distal to the tumor to facilitate esophageal dissection. Dissection is continued in an inferior direction so that the hiatus muscle of the diaphragm can be exposed and the cardia can be delivered into the thoracic cavity. During this procedure the lower portion of the periesophageal nodes and the posterior mediastinal nodes are removed together with the lower esophagus (Fig. 12).

Dissection of the Esophagus and the Periesophageal Nodes

Following the completion of the lower esophageal dissection, a Petz stapler is applied to transect the esophagus. This stapler is easy to manipulate and prevents contamination in the operative field. By pulling up the distal esophageal stump, the esophagus is dissected from the aorta and the pericardium (Fig. 13). The esophageal vessels should then be doubly ligated and divided. For fine vessels, hemoclips are convenient. In cases of cervico-thoraco-abdominal dissection, the thoracic duct should be ligated in the lower mediastinum to prevent chylorrhea, and its proximal portion is then resected because of the occasional presence of tumor thrombi.

Fig. 12. Dissection of the lower posterior mediastinal nodes (from [4])

Fig. 13. Dissection of the esophagus and the periesophageal nodes (from [4])

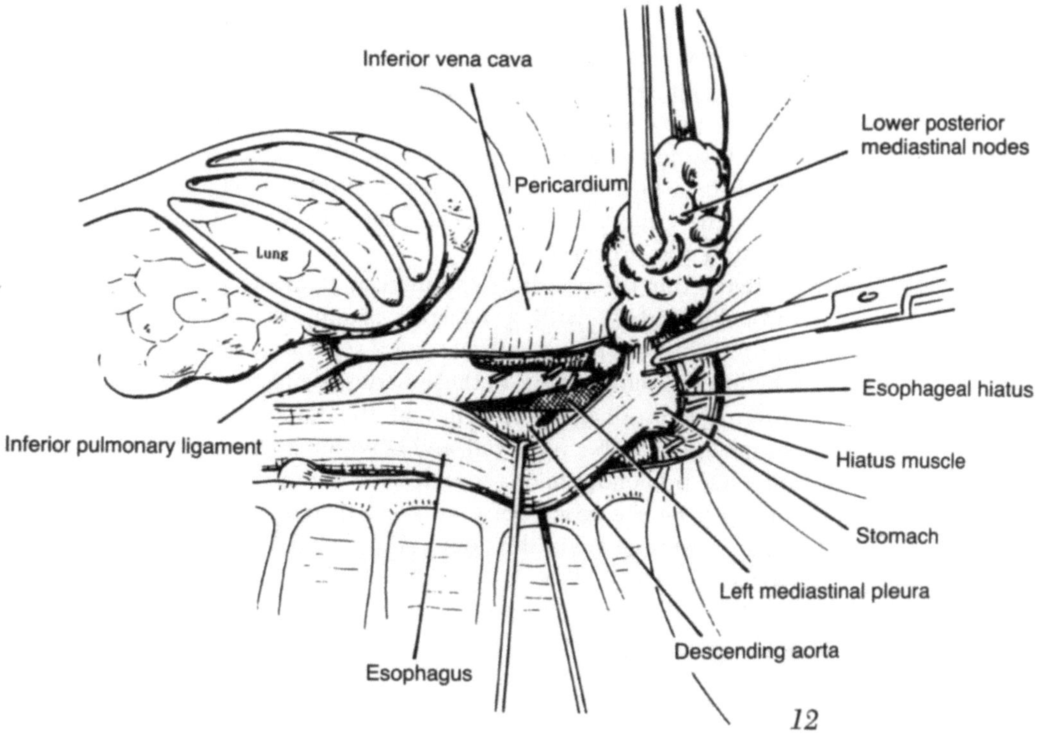
Inferior vena cava
Lower posterior mediastinal nodes
Pericardium
Lung
Esophageal hiatus
Inferior pulmonary ligament
Hiatus muscle
Stomach
Left mediastinal pleura
Descending aorta
Esophagus

12

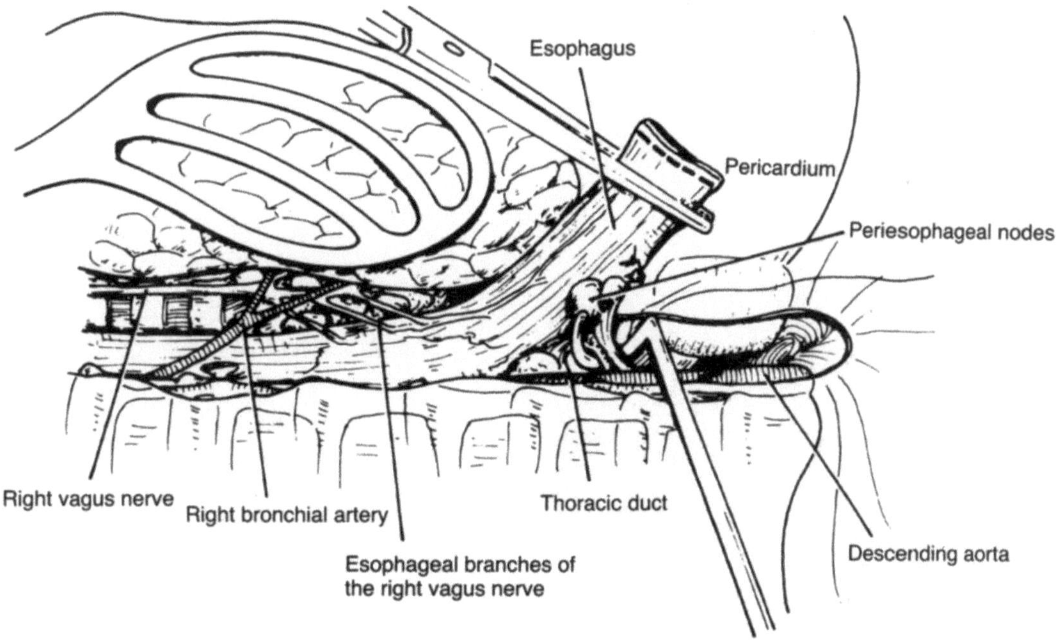
Esophagus
Pericardium
Periesophageal nodes
Right vagus nerve
Right bronchial artery
Thoracic duct
Descending aorta
Esophageal branches of the right vagus nerve

13

Dissection of the Infracarinal Nodes

The right bronchopulmonary nodes are present among the network of the pulmonary (bronchial) branches of the vagus nerve and the bronchial artery, posterior to the right main bronchus. These nodes are picked up with a clamp and removed using an ultrasonic aspirator in order to avoid damage to the nerve and the artery (Fig. 14). The esophageal branches of the right vagus nerve are divided after its lower bronchial branch is safely isolated and preserved.

Retracting the preserved pulmonary branches of the vagus nerve and the bronchial artery with a Babcock clamp, the infracarinal nodes situated posterior to the nerve and artery are separated from the carina. Branches of the left bronchial artery to these nodes are ligated with hemoclips (Fig. 15). After the esophageal branches of the left vagus nerve are divided, the esophagus can be retracted and a wider operative field can be obtained. The left bronchopulmonary nodes are then pulled out with scissors, using a clamp from the space between the left main bronchus and the descending aorta, and removal of the nodes is performed as far as the left margin of the aorta.

Fig. 14. Dissection of the right bronchopulmonary nodes (from [4])

Fig. 15. Dissection of the infracarinal nodes (from [4])

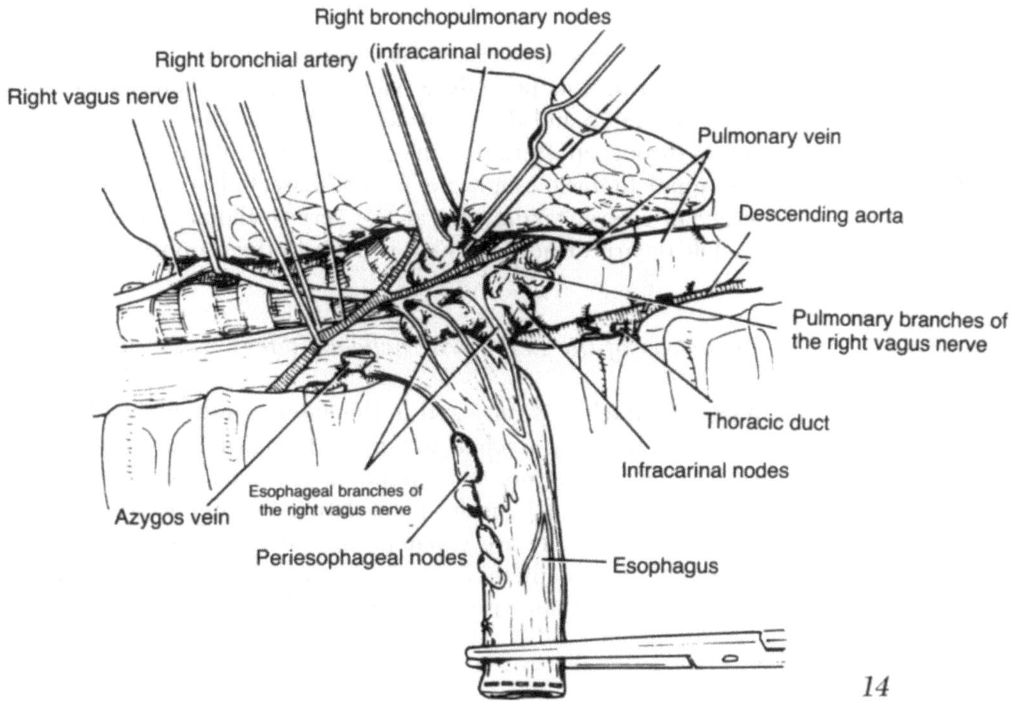
Right bronchopulmonary nodes
(infracarinal nodes)
Right bronchial artery
Right vagus nerve
Pulmonary vein
Descending aorta
Pulmonary branches of
the right vagus nerve
Thoracic duct
Infracarinal nodes
Azygos vein
Esophageal branches of
the right vagus nerve
Periesophageal nodes
Esophagus

14

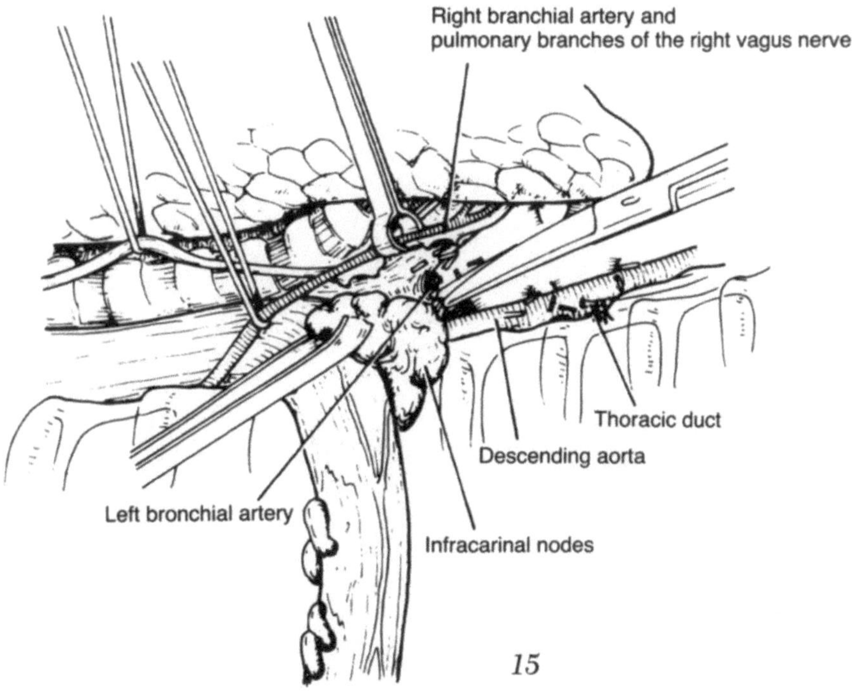
Right branchial artery and
pulmonary branches of the right vagus nerve
Thoracic duct
Descending aorta
Left bronchial artery
Infracarinal nodes

15

Dissection of the Upper Thoracic Esophagus

The upper thoracic esophagus is isolated from the tracheal membrane and taped. During this procedure, care should be taken not to injure the left recurrent nerve and the paratracheal sheath. The lower esophagus is then pulled through posterior to the preserved right bronchial artery (Fig. 16). The upper thoracic esophagus is then dissected upwards to the level of the suclavian artery, taking care at the cervico-thoracic junction not to injure the left recurrent nerve, which is close to the esophagus.

Dissection of the Left Paratracheal Nodes

While the trachea is retracted to the anterior side by hand, the fatty tissue is isolated from the trachea and divided longitudinally near the aortic arch. The left recurrent nerve is found in the fatty tissue and preserved; an ultrasonic aspirator is useful for isolating this nerve. The left paratracheal nodes which usually are situated anterior to the nerve are then removed following the complete isolation of the recurrent nerve (Fig. 17). The thoracic duct is then resected and ligated at the cervico-thoracic junction.

Fig. 16. The lower esophagus is pulled through posterior to the preserved right bronchial artery (from [4])

Fig. 17. Dissection of the left paratracheal nodes (from [4])

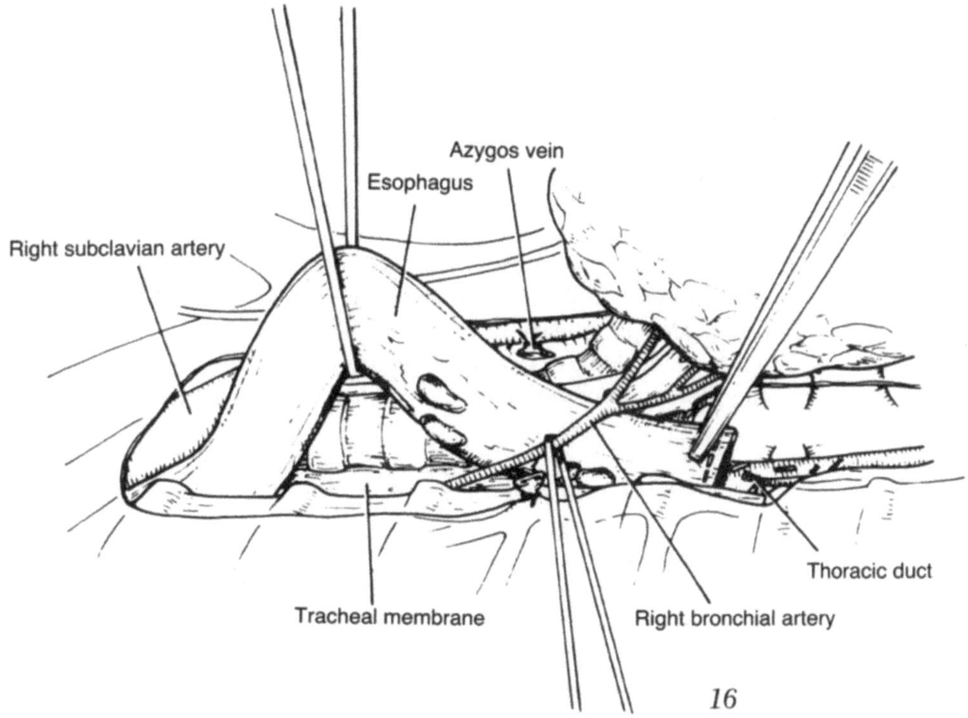
Azygos vein
Esophagus
Right subclavian artery
Thoracic duct
Tracheal membrane
Right bronchial artery

16

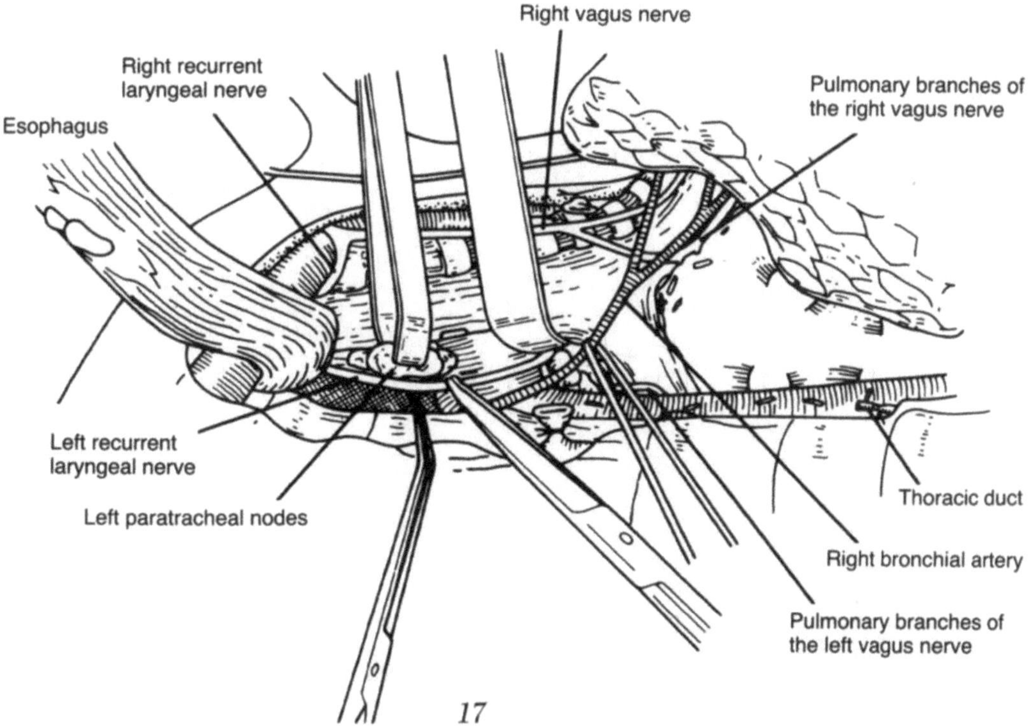
Right vagus nerve
Right recurrent laryngeal nerve
Pulmonary branches of the right vagus nerve
Esophagus
Left recurrent laryngeal nerve
Left paratracheal nodes
Thoracic duct
Right bronchial artery
Pulmonary branches of the left vagus nerve

17

Dissection of the Infraaortic Arch Nodes

The left recurrent nerve must first be isolated completely to the aortic arch. Retracting the left pulmonary artery, the infraaortic nodes are then removed using an ultrasonic aspirator (Fig. 18). The left bronchial arteries from the aortic arch should be preserved as much as is possible. The dissection margin in this procedure is as far as the mediastinal pleura of the left side.

Figure 19 illustrates the situation following esophagectomy and mediastinal lymph node dissection with preservation of the right bronchial artery and pulmonary branches of the vagus nerve.

After checking to confirm that there is no bleeding and no lymphorrhea, the thoracotomy wound is closed, leaving a drainage tube.

Laparotomy and Isolation of the Stomach

For laparotomy the patient is first placed in the supine position. The greater omentum and the gastrolienal ligament are divided between ligations, preserving the right and left gastroepiploic arteries (Fig. 20). When the posterior gastric vessels are ligated and divided, the posterior wall of the stomach is then sufficiently isolated.

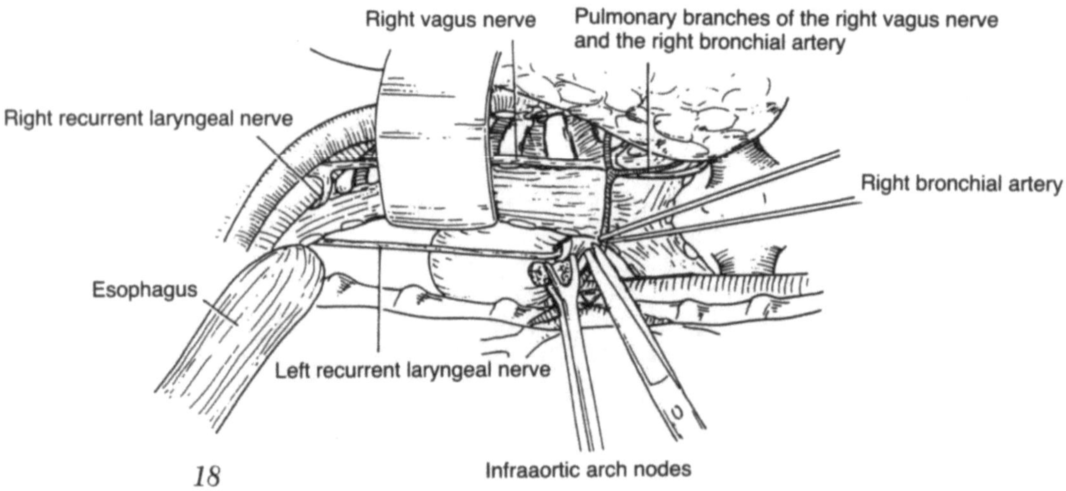

18

Fig. 18. Dissection of the infraaortic arch nodes (from [4])

Fig. 19. Functional mediastinal lymph node dissection preserving the right bronchial artery and pulmonary branches of the right vagus nerve (from [4])

Fig. 20. Isolation of the stomach, dividing the greater omentum and the gastrolienal ligament (from [4])

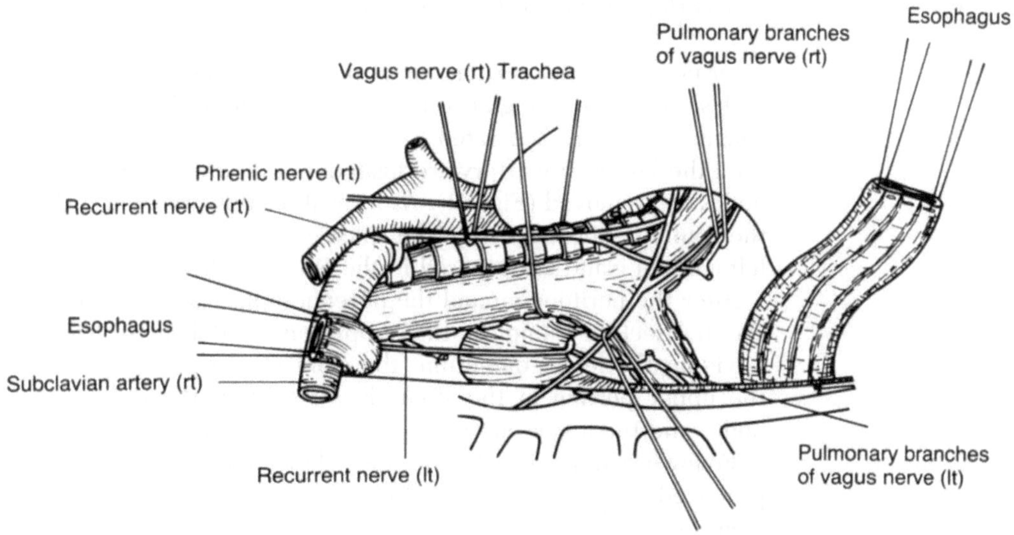
Esophagus
Pulmonary branches
of vagus nerve (rt)
Vagus nerve (rt) Trachea
Phrenic nerve (rt)
Recurrent nerve (rt)
Esophagus
Subclavian artery (rt)
Recurrent nerve (lt)
Pulmonary branches
of vagus nerve (lt)
Bronchial artery (rt)

19

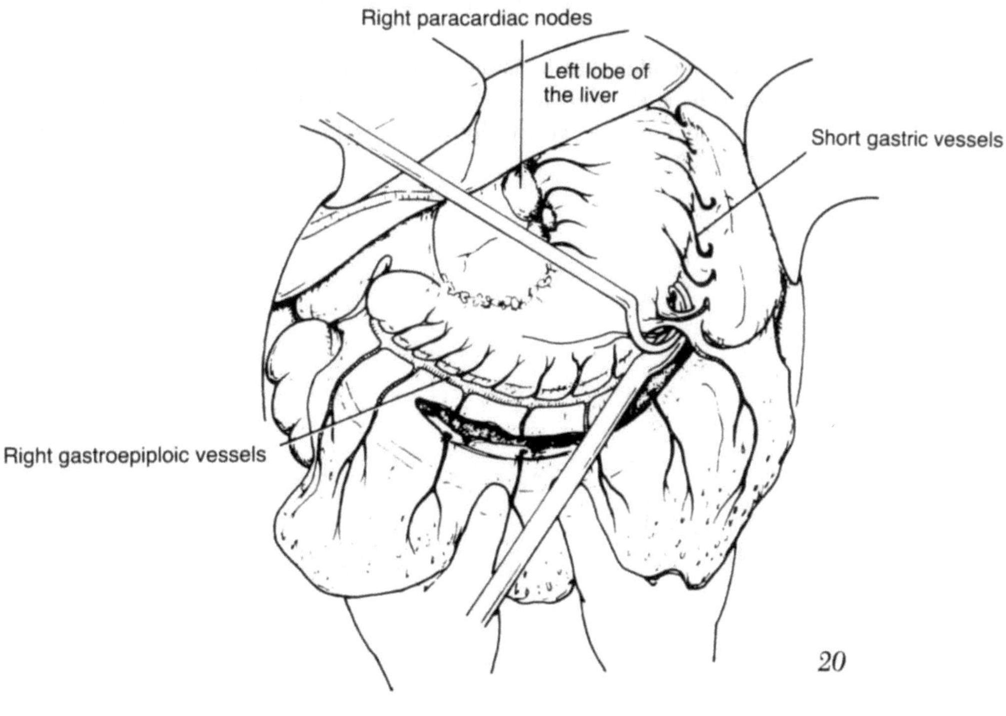
Right paracardiac nodes
Left lobe of
the liver
Short gastric vessels
Right gastroepiploic vessels

20

Dissection of the Abdominal Nodes

After incision of the peritoneum at the superior margin of the pancreas, the nodes along the left gastric artery, the common hepatic artery, and the splenic artery are separated from their vessels. The left gastric vein and then the left gastric artery are ligated and divided. Next, the celiac artery nodes are removed (Fig. 21). As a result of this procedure, the celiac axis, the common hepatic artery, the splenic artery, the hiatus muscle, and the left adrenal gland are exposed in the operative field.

After the peritoneum and the phrenoesophageal membrane are divided, the transected stump of the esophagus is delivered from the hiatus. Dividing the lesser omentum, the stomach is completely isolated. When the upper portion of the stomach is resected, the right and left cardial nodes and the lesser curvature nodes are simultaneously removed.

Following abdominal lymph node dissection, a pedicled gastric tube is produced by using a linear cutter such as GIA (US Surgical Co.) or PLC (Johnson and Johnson Co.). Pyloroplasty is performed by the "finger-crush" method.

Skin Incision and Division of the Anterior Cervical Muscles

After a collar skin incision is made in the neck, the left sternocleidomastoid muscle is divided near the sternum and the clavicle. The left sternohyoid muscle and the sternothyroid muscle are also divided near the sternum (Fig. 22).

Fig. 21. Dissection of the abdominal lymph nodes (from [4])

Fig. 22. Division of the anterior cervical muscles (from [4])

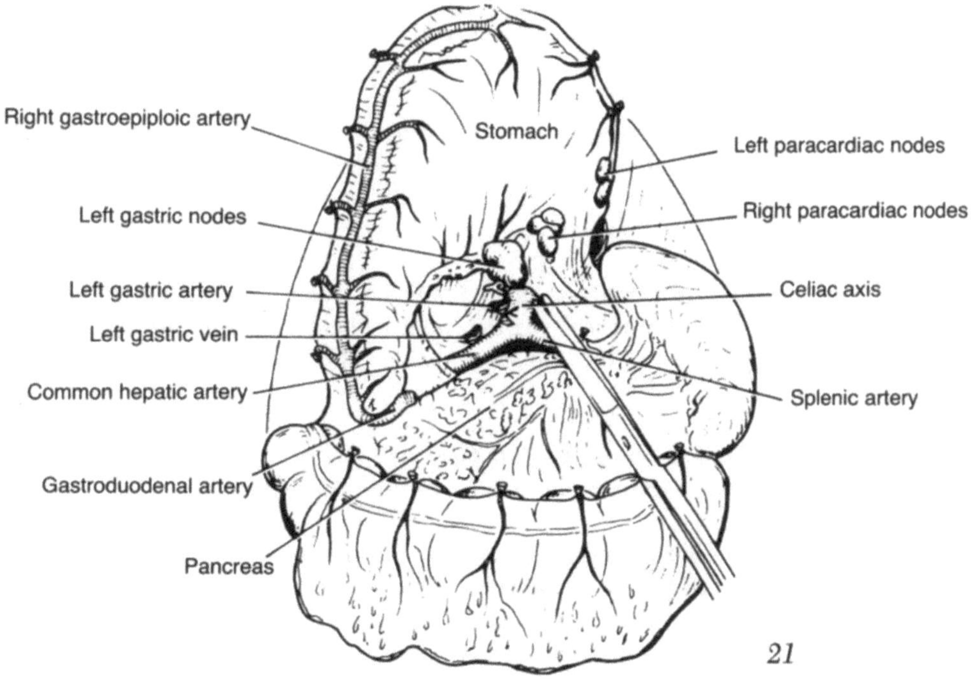
Right gastroepiploic artery
Stomach
Left paracardiac nodes
Left gastric nodes
Right paracardiac nodes
Left gastric artery
Celiac axis
Left gastric vein
Common hepatic artery
Splenic artery
Gastroduodenal artery
Pancreas

21

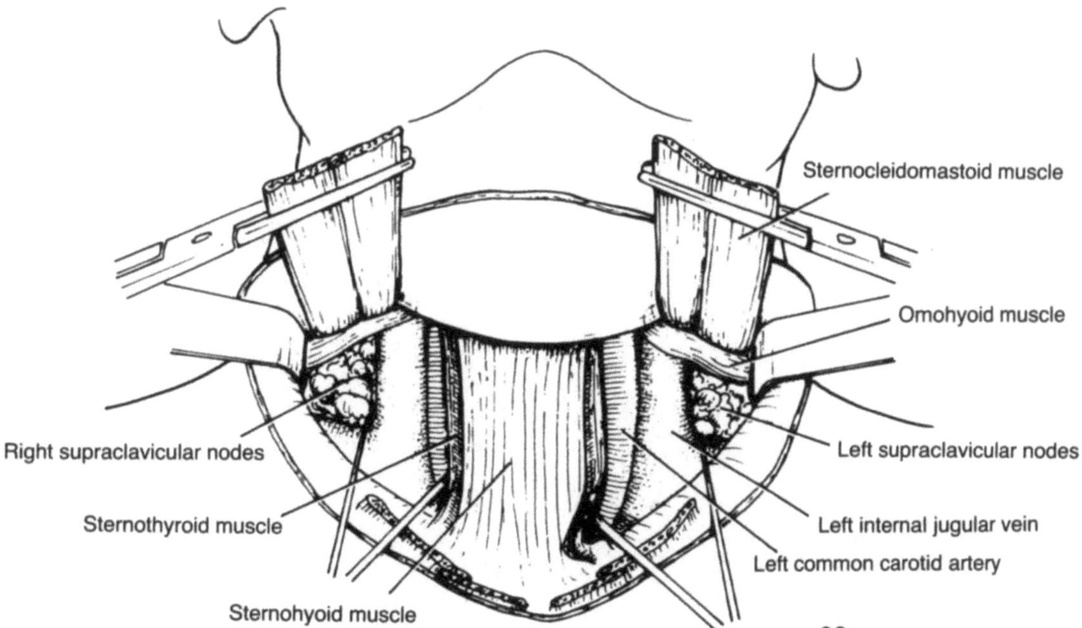
Sternocleidomastoid muscle
Omohyoid muscle
Right supraclavicular nodes
Left supraclavicular nodes
Sternothyroid muscle
Left internal jugular vein
Left common carotid artery
Sternohyoid muscle

22

Dissection of the Left Supraclavicular Nodes (Deep Lateral Cervical Nodes) and the Internal Jugular Nodes (Deep Anterior Cervical Nodes)

En-bloc resection is performed in the field surrounded by the left subclavian vein, the left internal jugular vein, and the omohyoid muscle. The deep cervical fascia (lamina prevertebralis) is preserved to avoid any injury to the phrenic nerve. The thoracic duct is ligated when it is found (Fig. 23).

Next, while the left omohyoid muscle is retracted to the superior side, the left internal jugular nodes along the left internal jugular vein are removed to the superior portion, as far as possible.

Dissection of the Left Cervical Paraesophageal and Paratracheal Nodes (Recurrent Nerve Lymphatic Chain)

The left internal jugular vein and the left common carotid artery are taped together and retracted to the left side. When the trachea is held to the right side with a retractor, the left recurrent nerve is then found in the tracheoesophageal groove. The cervical paraesophageal and paratracheal nodes are removed, avoiding any injury to the recurrent nerve and the left inferior thyroid artery. The paratracheal nodes are only present close to the thyroid. However, the greater portion of the cervical nodes between the trachea and the left common carotid artery has already been removed during the thoracic procedure. The esophagus is moved out into the neck, taking care to preserve the recurrent nerve. The cervical paraesophageal nodes, if present, are then removed (See Fig. 24).

Fig. 23. Dissection of the left supraclavicular nodes and internal jugular nodes (from [23])

Fig. 24. Bilateral modified neck dissection for carcinoma of the thoracic esophagus. *JV*, left jugular vein; *CA*, left carotid artery; *LRN*, left recurrent pharyngeal nerve

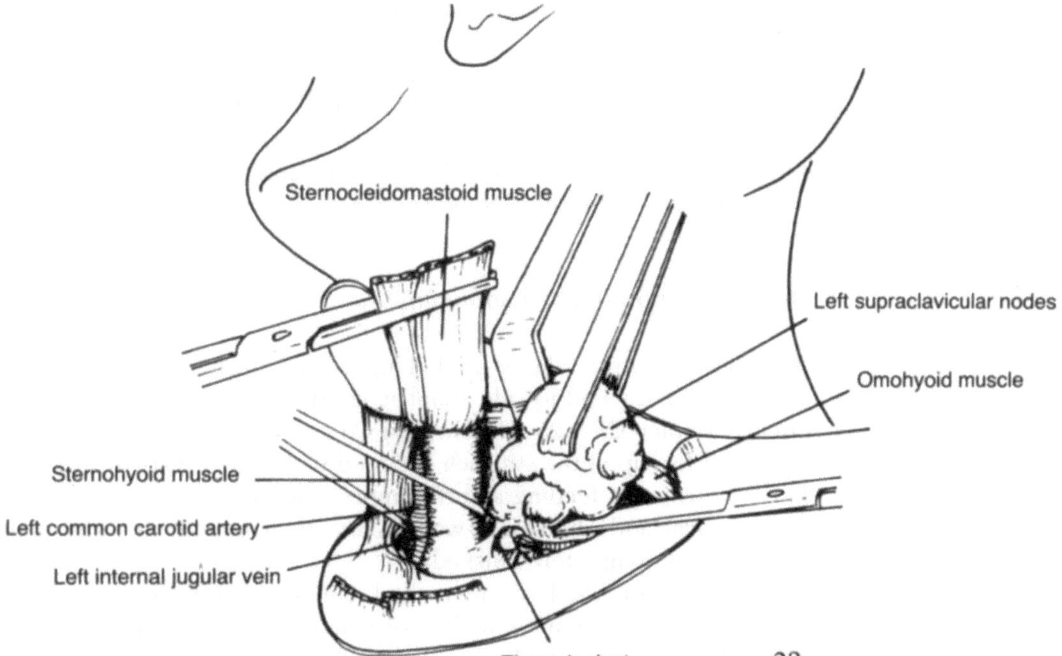
Sternocleidomastoid muscle
Left supraclavicular nodes
Omohyoid muscle
Sternohyoid muscle
Left common carotid artery
Left internal jugular vein
Thoracic duct

23

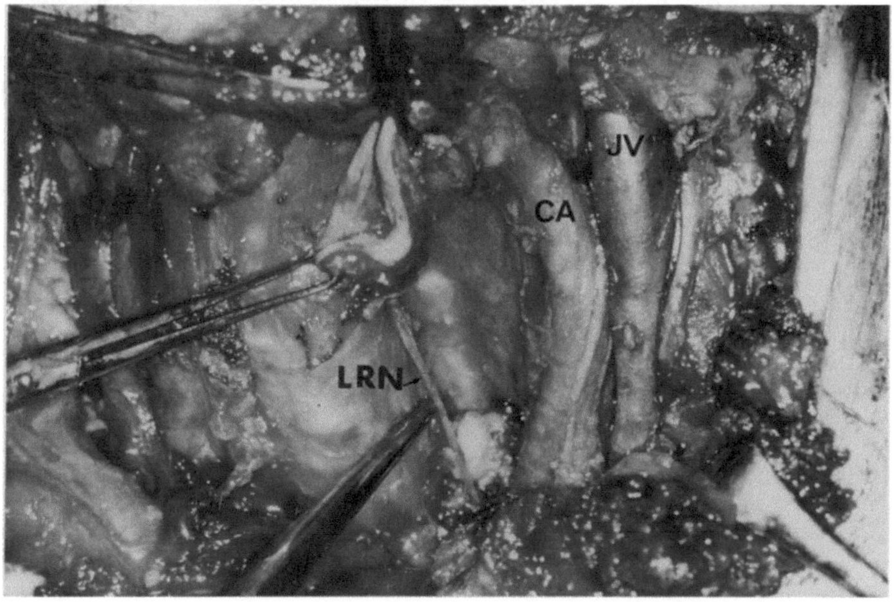
JV
CA
LRN

Dissection of the Right Cervical Lymph Nodes

The right cervical nodes are removed in the manner. The sternohyoid and sternothyroid muscles are not divided on the right side. The right paratracheal nodes anterior to the right recurrent nerve must then be removed (Fig. 25). However, the nodes posterior to the nerve are already removed during the thoracic procedure.

Reconstruction of the Esophagus

A pedicled gastric tube is pulled up to the neck retrosternally or subcutaneously. An end-to-side esophagogastrostomy is performed using the circular stapler EEA 25 (US Surgical Co.) or ILS 25 (Johnson and Johnson Co.), and the tip of the stomach is closed with a GIA. Gastrostomy is made in the antrum for feeding and decompression of the gastric tube (Fig. 26). The cervical and abdominal wounds are then closed, and the operative procedure is completely finished.

In this paper we have presented our surgical technique of cervico-thoraco-abdominal lymph node dissection for carcinoma in the thoracic esophagus. We have had 40 patients who have undergone this operative procedure. There was no operative death and only one hospital death (2.5%). The survival curves for the cervico-thoraco-abdominal lymph node dissection group (n = 40) and the thoraco-abdominal lymph node dissection group (n = 100), based on Kaplan-Meier analysis, are shown in Fig. 27. Both groups underwent curative resection of the total thoracic esophagus through right thoracotomy. The 5-year survival rate after cervico-thoraco-abdominal lymph node dissection was 50%, while that after thoraco-abdominal lymph node dissection was only 20%. There was a significant difference between these two curves on the Wilcoxon test. From these results, we conclude that cervico-thoraco-abdominal lymph node dissection is the best approach for carcinoma in the thoracic esophagus.

Fig. 25. Dissection of the right cervical paratracheal nodes (from [4])

Fig. 26. Reconstruction of the esophagus, showing drainage tubes (from [26])

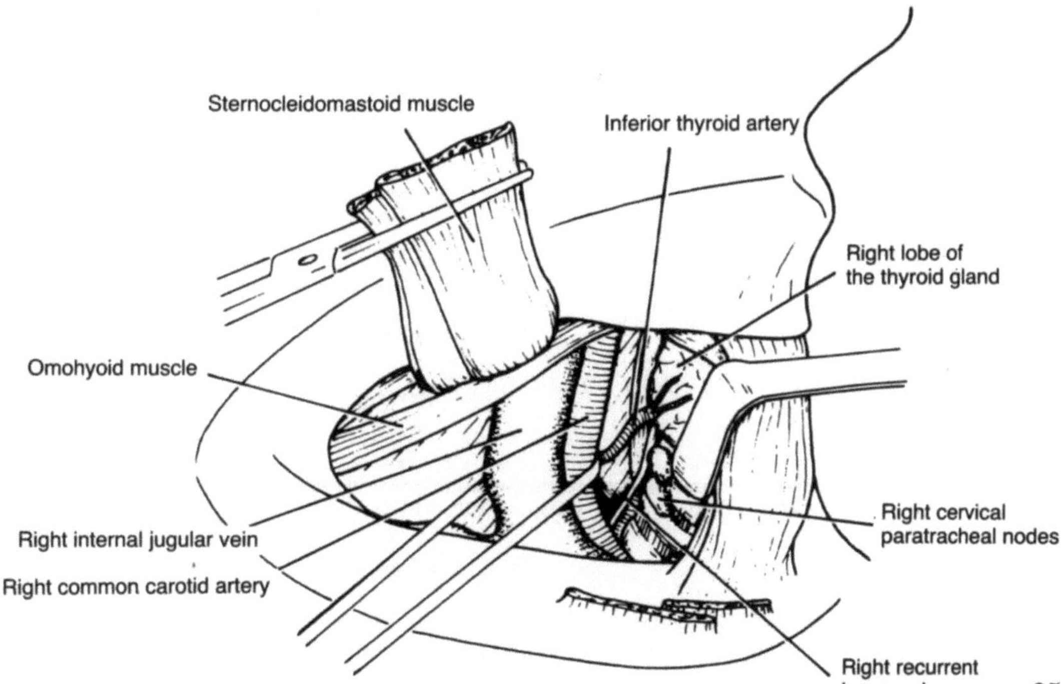
Sternocleidomastoid muscle
Inferior thyroid artery
Right lobe of the thyroid gland
Omohyoid muscle
Right internal jugular vein
Right common carotid artery
Right cervical paratracheal nodes
Right recurrent laryngeal nerve

25

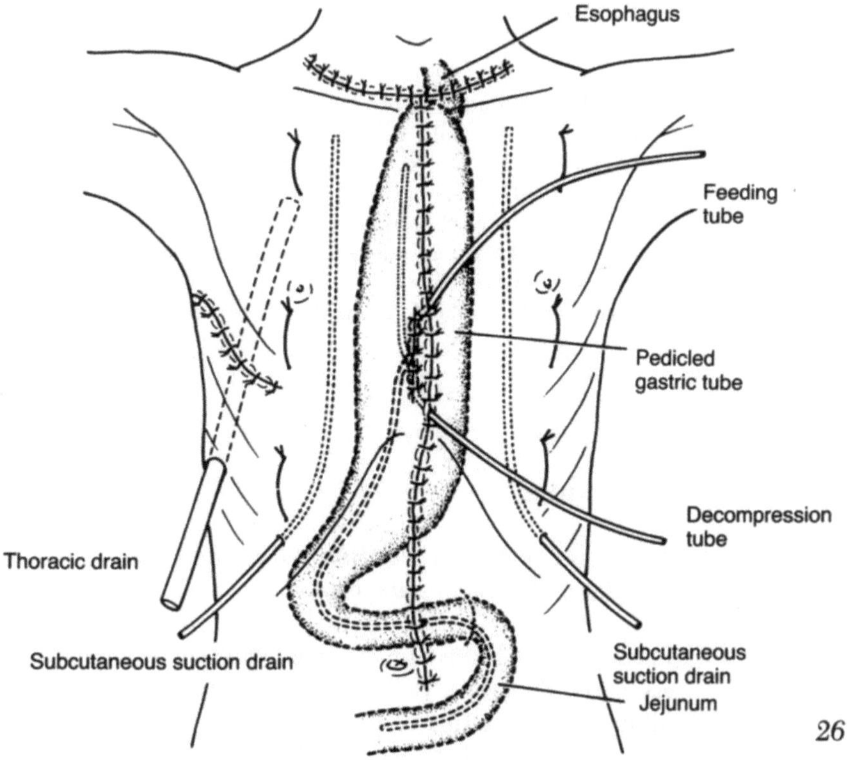
Esophagus
Feeding tube
Pedicled gastric tube
Decompression tube
Thoracic drain
Subcutaneous suction drain
Subcutaneous suction drain
Jejunum

26

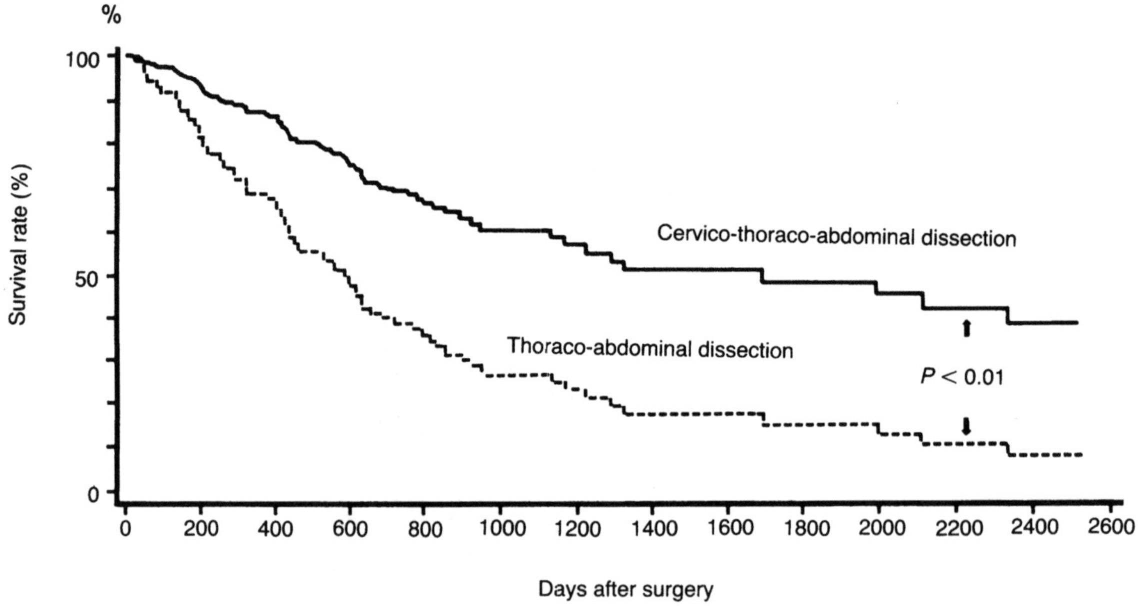

Fig. 27. Survival curves of the cervico-thoraco-abdominal lymph node dissection group vs the thoraco-abdominal lymph node dissection group, who were all curatively resected via right thoracotomy

References

1. Japanese Society for Esophageal Disease (1969) Guidelines for the clinical and pathological studies on carcinoma of the esophagus. Kanehara, Tokyo
2. Sannohe Y, Hiratsuka R, Doi K (1981) Lymph node metastasis in cancer of the thoracic esophagus. Am J Surg 141: 216–218
3. Sato T (1988) Presentation of illustration of surgical anatomy specially prepared for lymph node dissection of esophageal cancer. ISDE-Research Committee meeting on TNM-classification of esophageal carcinoma. Tokyo June 19. Book of Abstracts, pp 37–38
4. Kakegawa T, Fujita H, Yamana H (1991) Surgery for esophageal cancer. Igaku Shoin, Tokyo

Addendum

Methods for Accurately Recording the Dissected Lymph Nodes in Esophageal Cancer Surgery

Recording the status of metastatic lymph nodes in each individual operated case according to a predetermined modus operandi is a matter of great importance, but this can be rather troublesome if the procedure is not properly managed, which, in turn, would lead to insufficient records and data. Japanese surgical oncologists have over the years developed the following method for the successful systematic recording of dissected lymph nodes (Fig. 28).

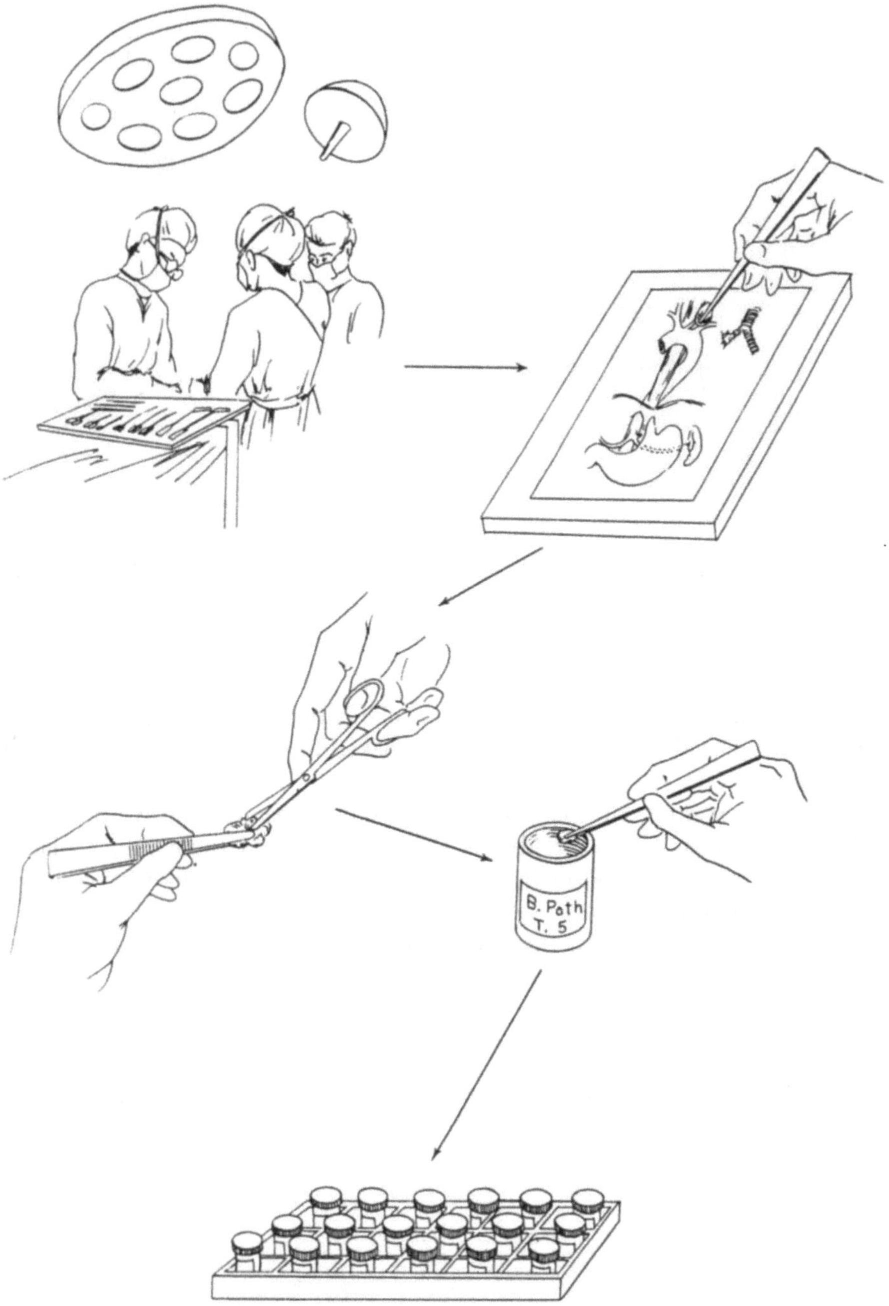

Fig. 28. Modus operandi for recording dissected lymph nodes

The surgery team must be accompanied by a para-surgical assistant, who notes the gross findings or macroscopic appearance that indicate the presence or absence of metastasis of the lymph nodes according to the remarks of the surgeons as the operation progresses.

Lymph nodes dissected during surgery are placed individually by the para-surgical assistant onto a flat schematic illustration map of the esophagus and the related nodes (Fig. 29). Nodes from the same location are therefore

Fig. 29. Para-surgical assistant places the dissected lymph node on a flat schematic map (ISDE News No. 8)

grouped together on the map. The flat schematic map on which the lymph nodes are placed is covered with gauze soaked in saline solution to prevent drying out.

In the next step, the para-surgical assistant separates the lymph nodes from the attached fatty or fibrous tissue, and places them into a plastic bottle containing 10% formalin, which is labeled with the location number of the lymph node.

The plastic bottles are then delivered to the pathology laboratory for histologic examination.

The Fourth Edition of the UICC TNM Classification of Esophageal Carcinoma and Its Relevance for Comparison of International Data

Toshifumi Iizuka[1]

Introduction

The TNM classification of the esophagus first appeared in the 1968 edition of the TNM *Classification of Malignant Tumours* [1]. In this edition, the regions, the extent of the primary tumor (T), regional lymph nodes (N), and distant metastases (M) classification of the esophagus was defined. This classification was made in 1966, with a trial period from 1967 to 1971. The TNM classification was revised and the second edition appeared in 1974 [2]. The revision was based on a proposal from the United States of America. This proposal was based on an analysis of more than 1000 patients with esophageal cancer that was conducted by the Task Force on the Esophagus of the American Joint Committee for Cancer Staging and End Results Reporting. The American system used three types of evaluation systems for classifying the extent of the disease, namely clinical, surgical evaluative, and post-surgical treatment classifications. Staging groups were defined in the second edition. In the third edition, published in 1978 [3], staging was changed from three to four groups, but the regions and the T, N, and M classification remained unchanged.

Because of some points which needed to be revised in the third edition, the Japanese Committee for Registration of Esophageal Carcinoma proposed a new TNM classification in 1985; this was based on an analysis of 3681 patients registered in Japan from 1969 to 1978. This proposal was accepted by the International Union Against Cancer (UICC), and the fourth edition, revised in accordance with the Japanese proposal, was published in 1987 [4].

In this chapter, the classifications of the third and fourth edition are examined, and the data for Japanese patients with esophageal carcinoma, which provided the basis of the revision of the fourth edition, are presented.

[1] National Oji Hospital, Kita-ku, Tokyo, 115 Japan

Table 1. Main points of difference between the third and fourth editions of the TNM classification.

	Third edition	*Fourth edition*
Region; demarcating points	Lower border of thoracic vertebra	Tracheal bifurcation
T; classification criteria	Length of tumor, invasion to surrounding structures	Depth of invasion
N for thoracic esophagus	Mediastinal	Mediastinal and perigastric

Main Revised Points of the Fourth Edition

The fourth edition included many revisions of the subject matter in the third edition. As seen in Table 1, the landmarks dividing the four regions in the third edition changed from the lower borders of the thoracic vertebrae to the thoracic inlet and tracheal bifurcation.

In the third edition, T was classified according to tumor length, circumferential invasion, and invasion to the surrounding tissues, but in the fourth edition, T was also classified according to the depth of invasion. Regional nodes included only mediastinal nodes in the third edition, but mediastinal and perigastric nodes were included in the fourth edition.

Patients Registered in Japan from 1969 to 1978

The revisions in the fourth edition were based on patients registered from 1969 to 1978 in Japan. A total of 7742 esophageal carcinoma patients were registered from 181 hospitals in that period. Of those, 4376 patients from 160 hospitals underwent resection.

For evaluation of the survival rate, meticulous follow-up of patients is necessary. Institutes with a 5-year follow-up rate of over 90% were used to compute the survival rate in this study. Accordingly, 435 patients from 41 hospitals were excluded, because the 5-year follow-up rate was less than 90% in these institutes. Thus 3941 patients from 119 hospitals were used for the calculation of the survival rate. Carcinoma was present in the thoracic esophagus in 3681 of these patients. Excluding 470 patients for whom pathology description was incomplete, 3211 patients were included in this study, and the 5-year follow-up rate for these patients was 95%.

Patients' data were input into a HITAC M 160-II computer (Hitachi Ltd.). The survival rate was calculated by Cutler's actuarial method [5]. Significant differences were determined to be in the range of 95% confidence limits by Greenwood's formula.

Anatomical Regions and Sites

The upper margin of the esophagus was defined as the pharyngo-esophageal junction in the third edition, but this part was included in the hypopharynx (postcricoid area) in the oropharyngeal classification. In the fourth edition, the upper margin was defined as the lower border of the cricoid cartilage in accordance with the definition of the hypopharynx.

Table 2. Anatomical subsites.

Third edition (1978)

Cervical esophagus: This commences at the pharyngo-esophageal junction and ends at the thoracic inlet, approximately 18 cm from the upper incisor teeth

Intrathoracic esophagus (excluding lower esophagus)

i) The upper thoracic portion, extending from the thoracic inlet to the lower border of the sixth thoracic vertebra, approximately 26 cm from the upper incisors

ii) The mid-thoracic portion, extending from the lower border of the sixth thoracic vertebra to the lower border of the eighth thoracic vertebra, approximately 31 cm from the upper incisors

Lower esophagus

This portion, 10 cm in length, commences at the lower border of the eighth thoracic vertebra and ends at the cardiac orifice, approximately 40 cm from the upper incisor teeth

Fourth Edition (1987)

Cervical esophagus: This commences at the lower border of the cricoid cartilage and ends at the thoracic inlet (substernal notch), approximately 18 cm from the upper incisor teeth

Intrathoracic esophagus

i) The upper thoracic portion extending from the thoracic inlet to the level of the tracheal bifurcation, approximately 24 cm from the upper incisor teeth

ii) The mid-thoracic portion is the proximal half of the esophaus between tracheal bifurcation and the esophagogastric junction. The lower level is approximately 32 cm from the upper incisors

iii) The lower thoracic portion, approximately 18 cm in length (excluding abdominal esophagus), is the distal half of the esophagaus between the tracheal bifurcation and the esophagogastric junction. The lower level is approximately 40 cm from the upper incisors

The third and fourth edition definitions of the anatomical regions are shown in Table 2. There are four regions in each edition. There is no problem in dividing the esophagus into four regions, but there were problems concerning where to divide the thoracic esophagus.

In the third edition, landmarks of the intrathoracic esophagus were defined as the lower border of the sixth thoracic vertebra for the upper and middle thoracic esophagus, and the lower border of the eighth thoracic vertebra for the middle and lower thoracic esophagus. However, unless we use large X-ray films covering the area from the cervical to the lumbar vertebrae, it is not always easy to determine the number of thoracic vertebra on esophagrams. Since the tracheal bifurcation, in contrast, is easily detected on esophagrams, the carina is a better landmark to separate the upper and middle thoracic esophagus.

For the demarcation point between the middle and lower thoracic esophagus, it is easy to find the midpoint between the carina and the esophagogastric junction on esophagrams. There is a slight difference between the third and fourth editions in the length from the upper incisors for each dividing border.

Table 3. Definition of T—Primary tumor.

Third edition (1978)

Tis: Pre-invasive carcinoma (carcinoma in situ)

T0: No evidence of tumor

T1: Tumor which involves 5 cm or less of esophageal length, which causes no obstruction, which does not involve the entire circumference of the esophagus and which shows no evidence of extra-esophageal spread

T2: Tumor which involves more than 5 cm of the esophageal length and which shows no evidence of extra-esophageal spread or tumor of any size which produces obstruction and/or involves the entire circumference of the esophagus and which shows no evidence of extra-esophageal spread

T3: Tumor with evidence of extra-esophageal spread

TX: The minimum requirement to assess the primary tumor cannot be met

Fourth edition (1987)

TX: Primary tumor cannot be assessed

T0: No evidence of primary tumor

Tis: Carcinoma in situ

T1: Tumor invades lamina propria or submucosa

T2: Tumor invades muscularis mucosa

T3: Tumor invades adventitia

T4: Tumor invades adjacent structures

Table 4. T classification (3rd edition) in relation to survival rate (%).

	No. of patients	*1*	*2*	*3*	*4*	*5 Years*
T0	50	50.9	31.2	31.8	32.6	33.3
T1	651	69.2	45.5	34.3	31.7	29.2
T2	1722	57.7	36.6	30.3	26.1	24.0
T3	545	48.9	24.7	17.0	14.0	12.9
TIS	17	42.0	24.5	25.0	19.2	19.6
Total	3012	58.4	36.3	28.7	25.1	23.2

Standard error of 5-year survival rate.

Total (1.67), T0 (32.20), T1 (3.92), T2 (2.24), T3 (3.05), Tis (20.58).

Figures in parentheses are percentages

T Classification

As seen in Table 3, the T classification in the third edition was based on the size of the tumor (i.e., whether it was 5 cm, more than 5 cm, or less than 5 cm), tumor extent, and the presence of extra-esophageal invasion.

Table 4 shows the T classification according to the third edition in relation to the survival rate in patients registered in Japan. Approximately 57% of these patients had T2 disease. There were differences between the survival rates of the four tumor-bearing groups; there were statistically significant differences between T3 and T1 and T2, but there was no statistically significant difference between T1 and T2 patients.

To evaluate the prognostic reliability of the T-factor in the third edition, patients registered in Japan were examined in terms of the relationship of survival rate and length of tumor. Table 5 shows the survival rate according to the tumor length determined in 1-cm increments. The patients with tumors less than 2 cm in length showed the best survival rate, but there

Table 5. Length of tumor in relation to survival rate (%).

Length (cm)	*No. of patients*	*1*	*2*	*3*	*4*	*5 Years*
−0.9	16	75.4	55.8	49.7	50.7	37.0
1.0−	89	74.3	55.8	48.5	47.2	47.1
2.0−	299	66.8	49.1	40.1	36.0	29.3
3.0−	578	66.5	41.3	31.4	27.6	25.5
4.0−	667	62.7	35.6	28.3	24.8	23.7
5.0−	541	53.8	34.4	25.1	21.4	19.6
6.0−	337	51.9	32.6	26.6	21.7	20.5
7.0−	201	49.7	28.0	23.3	22.8	23.4
8.0−	102	38.0	19.5	18.9	15.0	14.3
9.0−	54	53.3	25.4	20.2	12.5	8.6
10.0−	37	27.7	8.5	8.8	6.0	3.1
11.0−	31	26.4	16.9	10.4	7.1	7.3
Unknown	60	44.9	29.3	24.3	21.1	19.5

Standard error of 5-year survival rate.
−0.9 (26.83), 1.0− (11.79), 2.0− (5.73), 3.0− (3.96), 4.0− (3.58), 5.0− (3.70), 6.0− (4.74), 7.0− (6.42), 8.0− (7.42), 9.0− (8.27), 10.0− (6.10), 11.0− (9.95), Unknown (11.21)

Table 6. Depth of invasion correlated with survival rate (%).

Depth	*No. of patients*	*1*	*2*	*3*	*4*	*5 Years*
EP	12	68.6	52.9	54.6	56.3	58.3
MM	18	83.0	52.2	53.4	54.8	56.3
SM	187	78.4	62.4	53.2	49.7	47.3
MP	607	65.5	46.0	37.9	34.0	29.5
A1	521	63.1	39.7	30.1	28.1	24.8
A2	1186	58.4	34.6	26.5	21.6	20.9
A3	483	35.7	13.8	10.3	8.5	8.0
Total	3012	58.4	36.3	28.7	25.1	23.2

Standard error of 5-year survival rate.
Total (1.67); EP, epithelium (33.64); MM, muscularis mucosa (28.16); SM, submucosa (8.17); MP, muscularis propria (4.06); A1, possible invasion to the adventitia (4.11); A2, definite invasion to adventitia (2.56); A3, invasion to the neighboring structures (2.63)

was no difference in survival rate among patients whose tumor size ranged from 2 to 8 cm in length. The survival rate dropped abruptly in patients with tumors more than 10 cm in length.

It is clear from Table 4 that there is no significant difference of survival rates depending on whether tumor length is less than 5 cm or not. Therefore, 5 cm is not a good benchmark of tumor classification.

Table 6 shows the correlation between the depth of invasion classified according to the Japanese Guidelines [6] and the survival rate. Because the number of cases of EP (epithelium) and MM (muscularis mucosa) cancer are limited, it is difficult to compare their survival rates with those of other groups. Nevertheless, there are statistically significant differences among SM (submucosa), MP (muscularis propria), A_1 (possible invasion to adventitia), A_2 (definite invasion to adventitia), and A_3 (invasion to neighboring tissues) cases, and between MP, A_2, and A_3, and between A_2 and A_3 cases. Approximately 40% of the patients belong to the A_3 category. From the results shown in Tables 4 and 5, it can be seen that the depth of

Table 7. T classification (4th edition) correlated with survival rate (%).

	No. of patients	*1*	*2*	*3*	*4*	*5 Years*
T1	215	78.2	61.1	53.3	50.5	48.5
T2	607	65.5	46.0	37.9	34.0	29.5
T3	1707	59.8	36.1	27.6	23.6	22.1
T4	483	35.7	13.8	10.3	8.5	8.0
Total	3012	58.4	36.3	28.7	25.1	23.2

Standard error of 5-year survival rate.
Total (1.67), T1 (7.65), T2 (4.06), T3 (2.18), T4 (2.63)

Table 8. Regional lymph nodes for the thoracic esophagus.

Third edition	The mediastinal nodes
Fourth edition	The mediastinal and perigastric nodes, excluding the celiac nodes

invasion is a better index than the length of the tumor in the T classification. Therefore, the depth of invasion was adopted as a criterion of the T classification in the fourth edition.

Table 7 shows the correlation between the T classification in the fourth edition and the survival rate. There are many patients in T3 and T4, which suggests that many patients who received resective surgery in Japan in this period were in the advanced stage. There are statistically significant differences between each group, and the T classification number generally reflects the patient's prognosis.

In the fourth edition, because clinical T has come to correspond to pathological T, it is easy to compare the clinical and pathological classifications. However, how to diagnose the depth of invasion accurately by clinical examination is a matter of controversy.

Recently, double contrast esophagraphy and esophagoscopy have improved the detection of even superficial lesions of the esophagus, and CT scans now indicate the invasion of the tumor into surrounding tissues, and the presence of lymph node enlargement.

Uematsu [7] reported that in 329 of 376 (88%) patients with resected esophageal carcinoma, the depth of invasion was diagnosed accurately by radiographic examination. Kumegawa [8] reported that superficial carcinomas were diagnosed accurately by means of esophagoscopy and ultrasonography through endoscopy. The author's opinion is that the depth of invasion can be diagnosed in over 80% of patients by esophagraphy using the double contrast method.

N Classification

Because metastases to lymph nodes beyond the regional nodes are classified as distant metastases (M), the definition of regional nodes is important. There was no problem in the N classification for cervical esophagus, but questions concerning the regional nodes in the thoracic esophagus remain. Table 8 shows the classification of the regional nodes for the intrathoracic esophagus in the third and fourth editions.

In the third edition, the only regional nodes for the intrathoracic esophagus are the mediastinal nodes. In such cases, metastases to the

Table 9. Lymph node metastasis, classified according to the Japanese Guidelines, correlated with survival rates (%).

	No. of patients	*1*	*2*	*3*	*4*	*5 Years*
N(−)	1185	73.9	55.9	47.2	42.7	39.5
N1(+)	285	58.9	36.4	26.6	22.2	20.3
N2(+)	853	55.2	28.2	20.4	17.0	15.9
N3(+)	459	38.7	14.7	9.9	7.7	6.4
N4(+)	230	29.2	7.8	4.0	2.5	2.1
Total	3012	58.4	36.3	28.7	25.1	23.2

Standard error of 5-year survival rate.
Total (1.67), N(−) (3.14), N1(+) (5.15), N2(+) (2.72), N3(+) (2.44), N4(+) (2.05)

Table 10. N classification (4th edition) in relation to survival rate (%).

	No. of patients	*1*	*2*	*3*	*4*	*5 Years*
N0	1185	73.9	55.9	47.2	42.7	39.5
N1	1138	56.1	30.3	21.9	18.3	17.0
M1(LYN)	689	35.5	12.4	7.9	6.0	5.0
Total	3012	58.4	36.3	28.7	25.1	23.2

Standard error of 5-year survival rate.
Total (1.67), N0 (3.14), N1 (82.42), M1(LYN) (1.78)

abdominal lymph nodes, for instance the paracardiac nodes, are regarded as distant metastases, M1. The Japanese *Guidelines for Clinical and Pathologic Studies on Carcinoma of the Esophagus* [6] divides the lymph nodes into four groups; N_1, paraesophageal lymph nodes; N_2, other mediastinal and perigastric lymph nodes; N_3, abdominal lymph nodes other than perigastric nodes; and N_4, more distant nodes, including cervical lymph nodes.

Lymph node metastases in patients registered in Japan were classified according to the Japanese Guidelines, and the correlations between lymph node metastasis and survival rates are shown in Table 9. There was a statistically significant difference between the survival of metastases-negative and -positive groups. In the positive groups, there was no statistically significant difference between N1(+) and N2(+), but there were statistically significant differences between N1(+) or N2(+) and N3(+) or N4(+). There were many 5-year survivors in the N2(+) group, which includes involvement of both mediastinal and perigastric lymph nodes, but there were only a few long-term survivors in the N3(+) and N4(+) groups. Therefore, if regional lymph nodes are defined as mediastinal nodes only, as in the third edition, many patients with metastases to perigastric nodes, who have a good prognosis, will be assigned to the M1 group. This is inappropriate, considering the prognosis of these patients.

Survival according to the N classification of the fourth edition is shown in Table 10, and N classifications in the third and fourth editions are compared in Table 11. There is a significant difference between N0 and N1, and between N1 and M1 (LYN). M1 (LYN) means metastasis to lymph nodes more distant than the regional nodes. As seen in the lowest line of Table 12, M1 (metastases to distant organs) has poorer survival rate than M1 (LYN), therefore patients with M1 (LYN) should be classified as N2 rather than M1 in the next revision of TNM classification.

Table 11. Definition of metastasis to regional lymph nodes for the thoracic esophagus.

Third edition
N0: No evidence of involvement of regional lymph nodes on surgical exploration or mediastinoscopy
N1: Evidence of involvement of regional lymph nodes on surgical exploration or mediastinoscopy
NX: It is not possible to assess the regional lymph nodes (i.e., exploratory operation not done)

Fourth edition
NX: Regional lymph nodes cannot be assessed
N0: No regional lymph node metastasis
N1: Regional lymph node metastasis

Table 12. Combination of TNM (4th edition) and survival rate (%).

	No. of patients	*1*	*2*	*3*	*4*	*5 Years*
T1N0M0	136	89.9	73.9	67.6	63.4	60.8
T1N1M0	51	65.6	45.6	31.3	32.2	30.8
T2N0M0	316	77.1	63.2	52.6	46.9	41.6
T2N1M0	184	57.8	33.3	29.3	27.5	23.1
T3N0M0	607	73.6	54.3	48.5	41.2	38.9
T3N1M0	678	61.1	33.9	23.1	18.4	17.4
T4N0M0	113	53.3	26.8	19.8	18.4	15.9
T4N1M0	192	37.7	14.6	11.6	8.5	8.6
M1(LYN)	644	37.4	13.3	8.5	6.4	5.3
M1	91	20.4	4.4	1.5	1.5	1.6
Total	3012	58.4	36.3	28.7	25.1	23.2

Standard error of 5-year survival rate.
T1N0 (9.59), T1N1 (14.58), T2N0 (6.14), T2N1 (6.80), T3N0 (4.36), T3N1 (3.16), T4N0 (7.37), T4N1 (4.28), M1(LYM) (1.90), M1 (3.05)

Table 13. Stage classification.

Third edition				*Fourth edition*			
Stage I	T1	N0	M0	Stage 0	Tis	N0	M0
Stage II	T2	N0	M0	Stage I	T1	N0	M0
Stage III	Any T	N1	M0	Stage IIa	T2	N0	M0
Stage IV	Any T	Any N	M1		T3	N0	M0
				Stage IIb	T1	N1	M0
					T2	N1	M0
				Stage III	T3	N1	M0
					T4	N0	M0
					T4	N1	M0
				Stage IV	Any T	Any N	M1

Next, the cervical lymph nodes are not included in the regional node classification for the thoracic esophagus in the fourth edition. In patients with cancer of the upper or middle thoracic esophagus, metastases to the cervical nodes are not rare, and patients with metastases to cervical nodes have a better survival rate than those with metastases to distant organs. Therefore, cervical nodes should be classified as N3 in the next revision.

M Classification

The M classification includes not only distant organ metastases but also metastases to lymph nodes more distant than the regional nodes. As seen in Table 12, there is a difference in survival between the M1 (LYM) and M1 groups, but it is not statistically significant. These two groups are classified in the same category in the fourth edition, but M1 (LYM) should be classified separately from the M classification, for instance as N2 and N3, in the next revision.

Stage Grouping

Stage grouping is a most important part of the TNM classification for predicting patient prognosis. Stage groupings in the third and fourth editions are shown in Table 13. In the third edition, the stage of thoracic esophageal cancer was divided into four groups, mainly according to T, and then according to N and M classification.

To prepare a reasonable staging classification, patients registered in Japan were divided according to various combinations of the T, N, and M classifications; their survival rates are presented in Table 12.

T1N0M0 has the most favorable survival rate, followed by T2N0M0 and T3N0M0. T1N1M0 and T2N1M0 belong to the next most favorable group, and T3N1M0, T4N0M0, and T4N1M0 constitute the next group. M1 (LYM) and M1 had the worst survival rate. These groups were designated as Stages I, IIa, IIb, III, and IV, and the new stage grouping is shown under the heading, Fourth edition in Table 13.

Ten-Year Survival of Patients with Esophageal Carcinoma Registered in Japan

To examine the clinical usefulness of the new TNM classification, the correlation of the 10-year survival of patients registered in Japan between 1969 and 1980 and this classification was studied.

In this 12-year period, 10113 patients were registered from 234 institutions in Japan and the results of 5481 resected patients in this series were examined. The survival rates of these patients were calculated by the life-table method of Cutler [5], using a HITAC M 160-H computer (Hitachi). Differences between survival rates were considered to be significant when they were in the range of 95% confidence limits according to Greenwood's [9] formula. Overall, the survival rate was 23.8% at 5 years and 18.7% at 10 years.

As seen in Table 14, females had a better survival rate than males for each survived year, and these differences were statistically significant.

Next, these patients were examined according to the TNM classification. Table 15 shows the relationship between T classification and survival rate. About half of the patients were T3, the most common of the resected groups, and next in order of magnitude was T2, followed by T4. T1 accounted for only a small number, suggesting that T1 cases, meaning tumors invading only the mucosa or submucosa, were not easily detected. The survival rate decreased gradually from T1 through T4. T1 had survival rates of 81.7% at 1 year, 49.9% at 5 years and 36.1% at 10 years. In

Table 14. Ten-year survival, by sex.

	No. of patients	*1*	2	3	4	5	*10 Years*
All	5071	57.7	37.1	29.6	25.8	23.8	18.7
Male	4133	56.0	35.1	27.7	23.9	21.7[a]	16.6[b]
Female	1007	65.1	45.7	37.9	33.9	32.5[a]	27.3[b]

Actuarial survival rate.
[a] $P < 0.005$
[b] $P < 0.025$

Table 15. Ten-year survival in relation to T classification.

	No. of patients	*1*	2	3	4	5	*10 Years*
All	5481	57.7	37.1	29.6	25.8	23.7	18.6
T1	376	81.7	67.3	59.1	53.6	49.9	36.1
T2	1019	67.4	47.6	38.8	35.2	31.0	24.8
T3	2794	59.5	36.8	28.3	24.2	22.4	17.8
T4	966	33.4	14.6	11.0	8.9	8.2	5.9
Unknown	213	50.6	34.8	29.9	27.2	27.2	22.5

Table 16. Ten-year survival in relation to lymph node metastasis.

	No. of patients	*1*	2	3	4	5	*10 Years*
All	5071	57.7	37.1	29.6	25.8	23.7	18.6
N0	1843	74.1	57.1	48.7	44.5	41.2	32.4
N1	1736	57.0	32.3	23.7	19.7	17.9	14.1
Unknown	517	45.1	28.1	21.7	17.7	16.9	15.4
(M LYM)[a]	1062	35.4	13.9	9.1	6.4	5.5	2.9

[a] Metastasis to lymph nodes beyond regional nodes

Table 17. Ten-year survival in relation to metastasis to distant organs.

	No. of patients	*1*	2	3	4	5	*10 Years*
All	5071	57.7	37.1	29.6	25.8	23.7	18.6
M0	3885	64.8	44.2	35.9	31.7	29.3	23.9
M1	1186	34.4	13.4	8.9	6.5	5.5	3.0

Table 18. Ten-year survival according to stage.

	No. of patients	*1*	2	3	4	5	*10 Years*
All	5071	57.7	37.1	29.6	25.8	23.7	18.6
Stage I	232	91.6	78.9	73.8	69.5	64.2	48.0
Stage IIA	1306	75.6	58.2	49.1	44.3	40.9	32.2
Stage IIB	336	62.6	40.2	31.6	28.1	24.7	18.6
Stage III	1484	56.4	31.1	22.5	18.5	17.2	13.8
Stage IV	1105	35.5	13.6	9.0	6.5	5.4	2.9
Unknown	111	47.0	31.7	24.3	20.9	19.8	18.1

Table 19. Patients registered with the ISDE Research Committee, 1988.

	Belgium	*India*	*Japan*	*PRC*
Location				
Cervical		6(3.0%)	30(5.6%)	
Upper thoracic		13(6.5%)	46(8.7%)	24(23.5%)
Middle thoracic	7(53.8%)	71(35.7%)	302(56.8%)	63(52.0%)
Lower thoracic	6(46.2%)	104(52.3%)	154(24.5%)	
Unknown		5(2.5%)		
Lymph Node Metastases				
Negative	3(23.1%)	64(32.2%)	209(39.3%)	51(50.0%)
Positive	10(76.9%)	128(64.3%)	321(60.3%)	51(50.0%)
Not cited		7(3.5%)	2(0.4%)	
Total	13(100%)	199(100%)	532(100%)	102(100%)

Belgium: Universitaire Zenkenhuizen, Leuven
India: Tata Memorial Hospital, Bombay
PRC: Cancer Institute Hospital, Beijing
Japan: Thirty-eight Institutions

contrast, T4 had survival rates of 33.4% at 1 year, 8.2% at 5 years, and 5.9% at 10 years. There were statistical differences between the survival rates for each category.

Table 16 shows the survival rates according to lymph node metastases. N0 had more than double the survival rate of N1 from 3 years onward, and the difference was statistically significant. M (LYM), i.e., metastasis to lymph nodes more distant than regional nodes, had a significantly lower survival rate than N1.

Table 17 indicates distant metastasis in relation to survival rate. There was a clear difference between M0 and M1, but some patients in the M1 group survived for more than 10 years. This is because M1 includes not only organ metastasis, but also metastasis to lymph nodes more distant than the regional nodes.

Table 18 shows the survival rates according to staging. Stage 1 had the best survival rates, 64.2% at 5 years and 48.0% at 10 years. There were statistical differences between each of the five groups, suggesting that this staging reflected survival well.

ISDE Research Committee for TNM Classification

The ISDE (International Society for Diseases of the Esophagus) established a Research Committee for TNM classification on July 19, 1988. The committee was set up to organize acceptance of this classification throughout the world, and to investigate the clinical usefulness of this classification for international comparison of the treatment results for esophageal cancer. For this purpose, a registration form was drawn up after vigorous discussion among specialists from throughout the world, and resected cases have subsequently been registered from many institutions in different countries. These patients will be examined to determine differences in demographic factors and survival rates for each country.

In 1988, 844 patients from four countries were registered, as seen in Table 19. In Belgium, Japan and the People's Republic of China, over 50% of the patients had cancer of the middle thoracic esophagus, but 52.3% of

Table 20. Patients registered with the ISDE Research Committee, 1989/1.

	Belgium	*India*	*Italy*	*Japan*	*PRC*	*FRG*	*Yugoslavia*
No. of patients	34	272	76	1365	268	160	29
Sex							
Male	79%	68%	82%	86%	76%	86%	83%
Female	21%	30%	17%	14%	24%	14%	17%
Unknown		2%	1%				
Mean Age	58	50	57	62	56	57	56
Location							
Cervical		2%	21%	5%		1%	
Upper th.	3%	6%	13%	10%	25%	36%	
Middle th.	47%	32%	32%	54%	54%	19%	83%
Lower th.	46%	50%	34%	29%	19%	44%	17%
Unknown		14%		2%	2%		

Belgium: Universitaire Ziekenhuizen, Leuven
India: Tata Memorial Hospital, Bombay
Italy: Degli Studi Di Padova, Padova
Japan: Thirty-eight Institutions
PRC: Cancer Institute Hospital, Beijing
FRG: Technical University of Munich, Munich
Yugoslavia: Univerzietski Klinici Centra, Belgrade

Indian cases were in the lower thoracic esophagus. This phenomenon is very interesting, and its cause should be studied. Lymph node metastases were present in 50%–76.9% of patients. These results indicate that over half of the patients who underwent resection surgery were in the advanced stage, and this was one of the causes for the poor results of treatment for esophageal cancer.

In 1989, 2204 patients from seven countries were registered, as shown in Table 20. There were more male than female patients in all countries, 68%–86% of the patients being male. The incidence in females was higher in China and India than in the other five countries.

The mean age in India was lower than in the other countries, and Japan had the highest mean age. The cause of the high incidence in younger age groups seen in India and China should be studied, in addition to their higher incidence rates in females.

In most countries, the location of the tumor was most commonly in the middle thoracic esophagus, but in 1989 cancer of the lower thoracic esophagus was commonest in India and Belgium (Table 20). India had the highest incidence of cancer of the lower esophagus in 1988 (Table 19).

Next, these patients were evaluated by the TNM classification of the fourth edition, and the results are shown in Table 21. Regarding T-classification, p-T3 was the most common in all countries, ranging from 37% to 86%.

In Japan, 22% of patients had T1 tumors, in which cancer invasion is limited to the mucosa or submucosa, but other countries had only 1%–13% of T1 cases. In Japan, the widespread application of endoscopic examination has increased the number of T1 lesions detected in recent years.

Concerning lymph node metastasis, about 60% of patients in Belgium, India, Italy, and Japan had metastases, whereas fewer than 50% of patients in Germany and China had lymph node metastases.

The rate of M1 ranged from 4% to 18%, which is a high value for patients undergoing resection surgery, but this is partly due to the fact that M1

Table 21. Patients registered with the ISDE Research Committee, 1989/2, TNM classification (4th edition).

	Belgium	*India*	*Italy*	*Japan*	*PRC*	*FRG*	*Yugo-slavia*
No. of patients	34	272	76	1365	268	160	29
Distribution of TNM classification							
p-T0		3%	3%	2%	0.4%		
p-T1	9%	1%	9%	22%	3%	13%	4%
p-T2	6%	5%	17%	18%	8%	16%	48%
p-T3	74%	86%	45%	37%	62%	46%	48%
p-T4	6%	2%	21%	16%	26%	21%	
Not cited	5%	3%	1%	2%			
N classification							
p-N0	32%	33%	34%	36%	57%	65%	45%
p-N1	68%	64%	66%	59%	42%	35%	55%
Not cited		3%		5%	1%		
M classification							
M0	76%	96%	86%	81%	96%	96%	48%
M1	18%	4%	12%	17%	3%	4%	
Not cited	6%		2%	2%	1%		52%
Stage classification							
0		4%	5%	4%	1%	1%	
I	6%	1%	4%	14%	3%	10%	
IIA	21%	28%	18%	13%	34%	7%	45%
IIB	8%	1%	14%	14%	3%	14%	
III	41%	58%	44%	33%	55%	64%	
IV	18%	3%	12%	17%	3%	4%	
Not cited	6%	5%	3%	5%	1%		

includes not only distant organ metastasis, but also metastasis to lymph nodes more distant than the regional nodes.

Regarding stage classification, stage 0 and stage 1 were seen in more than 10% of patients in Japan and Germany, but such cases were rare in India and China. Stage III was the most common in all countries.

From the data cited above, it can be recognized that many cases being treated surgically are actually at an advanced stage, so if we were to deal with patients at an earlier stage, we would probably obtain better treatment results. The most important point in the successful treatment of esophageal cancer is detection at an early stage.

ISDE Lymph Node Grouping

When we perform curative surgery on patients with esophageal cancer, it is essential to know which lymph nodes are invaded by cancer cells. Lymph nodes are classified into groups according to anatomical location. Recently, Professor Sato of Tokyo Medical and Dental University, at the request of the ISDE TNM Research Committee, made a new lymph node grouping map based on a precise study of lymph flow (see p. 55). Lymph nodes are divided into three regions, cervical, mediastinal, and abdominal, which are subdivided into smaller groups. To investigate metastatic rate according to lymph node groups, patients who had received extended lymph node dissection in the cervical, mediastinal, and abdominal regions

Table 22. Tumor location in relation to lymph node metastasis in patients registered in Japan, 1989.

	Upper *40 cases*	*Middle* *149 cases*	*Lower* *75 cases*	*Total* *264 cases*
Cervical nodes	11(28%)	43(29%)	17(23%)	71(27%)
Cervical paratracheal	3(8%)	18(12%)	9(12%)	30(11%)
Internal jugular (right)	4(10%)	10(7%)	7(9%)	21(8%)
Internal jugular (left)	3(8%)	13(9%)	5(7%)	21(8%)
Supraclavicular (right)	1(3%)	12(8%)	3(4%)	16(6%)
Supraclavicular (left)	2(5%)	16(11%)	4(5%)	22(8%)
Mediastinal nodes	26(65%)	85(57%)	37(49%)	148(56%)
Periesophageal	8(20%)	46(31%)	25(33%)	79(30%)
Right recurrent nerve	11(28%)	29(19%)	9(12%)	49(19%)
Right paratracheal	5(13%)	14(9%)	5(7%)	24(9%)
Infracarinal	4(10%)	25(17%)	9(12%)	38(14%)
Left paratracheal	8(20%)	26(17%)	2(3%)	35(14%)
Infraaortic Arch	4(10%)	4(3%)	2(3%)	10(4%)
Inferior posterior mediastinal	2(5%)	23(15%)	13(17%)	38(14%)
Abdominal nodes	4(10%)	60(40%)	47(63%)	111(42%)
Right paracardiac	2(5%)	32(21%)	29(39%)	63(24%)
Left paracardiac	2(5%)	21(14%)	22(29%)	45(17%)
Lesser curvature		15(10%)	14(19%)	29(11%)
Left gastric		24(16%)	22(29%)	46(17%)
Common hepatic		2(2%)	9(12%)	11(4%)
Splenic			3(4%)	3(1%)
Celiac		4(3%)	7(9%)	11(4%)
Abdominal aortic		1(1%)	7(9%)	8(3%)

Lymph node groups were classified according to the map of Professor Tatsuo Sato, Tokyo Medical and Dental University (p. 55)

were studied. For this purpose, 264 patients who had undergone such lymph node dissection were selected from the patients registered in Japan in 1989. Table 22 shows the correlation between location of the tumor and lymph node metastases. There were 40 patients with cancer of the upper thoracic esophagus. In this group, metastases were present not only in the mediastinal nodes, but also in the cervical nodes. The most frequent site of metastases was the right recurrent nerve nodes (28%), followed by the left paratracheal nodes and periesophageal nodes. In 28% of cases there was metastases to the cervical lymph nodes, of which the internal jugular nodes were most commonly involved.

In 149 patients with middle esophageal cancer, metastases were most common in the mediastinal, followed by abdominal, and then cervical lymph nodes. Most frequent sites were the periesophageal nodes, followed by the right recurrent nerve and right paracardiac nodes. The next most common groups were the infracarinal, left paratracheal, left paracardiac, left gastric, cervical paratracheal, and supraclavicular nodes. Cancer of the middle thoracic esophagus usually has widespread metastases to all three regions.

In 75 patients with lower thoracic esophageal cancer, a high frequency of metastases was seen in all three regions. The main metastatic sites were the periesophageal, right and left paracardiac, left gastric, right recurrent nerve, infracarinal, and cervical paratracheal nodes. In cancer of the lower esophagus, lymph node metastases were present mainly in the mediastinal and abdominal, followed by the cervical lymph nodes, suggesting that lymphatic channels from the mediastinum to the abdomen may play an important role.

From the observation of lymph node metastases in these patients with dissection in the three regions, it is clear that esophageal cancer metastasizes to lymph nodes not only near the main tumor but also to distant nodes. Of the mediastinal lymph node groups, the right recurrent nerve chain had the higher metastatic rate than other lymph nodes except periesophageal nodes, regardless of the tumor location. From these results, it can be seen that this lymph node group should be dissected in all cases of thoracic esphageal cancer.

Abdominal lymph node metastasis was frequently seen in both the middle and lower esophageal cancer. Metastatic rates were 40% for middle thoracic and 63% for lower thoracic cancer. In the abdominal lymph node groups, the paracardiac and left gastric lymph node groups had higher rates than the others. It is therefore necessary to dissect abdominal lymph nodes carefully in middle and lower esophageal cancer.

Metastasis to the cervical lymph nodes was similar in the three groups, being 28% for the upper thoracic, 29% for the middle thoracic, and 23% for the lower thoracic esophagus. In advanced esophageal cancer, cervical nodes should be dissected to obtain long-term survival.

Conclusion

The TNM classification has a long history, of more than 20 years, and it has been revised three times. For esophageal cancer, a fundamental revision of the T and N classification was made in 1987, according to the Japanese proposal based on the data of 3012 resected patients registered between 1969 and 1978. But problems remain concerning regional node classification. Abdominal lymph nodes other than perigastric nodes and cervical nodes are not included in regional nodes, therefore metastases to these lymph nodes are referred to as distant metastases (M). Because the survival rate of patients with metastases to these lymph nodes is better than that of patients with actual organ metastases, it is not reasonable to treat the two groups as one. These lymph nodes should be treated as regional nodes. These findings suggest that the regional node classification should be revised.

References

1. Committee on TNM Classification (1968) TNM classification of malignant tumors. International Union Against Cancer, Geneva, pp 25–26
2. Committee on TNM Classification (1974) TNM classification of malignant tumors, 2nd edn. International Union Against Cancer, Geneva, pp 39–42
3. Committee on TNM Classification (1978) TNM classification of malignant tumors, 3rd edn. International Union Against Cancer, Geneva, pp 57–61
4. Committee on TNM Classification (1987) TNM classification of malignant tumors, 4th fully revised edn. Springer, Berlin, pp 40–42
5. Cutler SJ (1964) Computation of survival rate. Natl Cancer Inst Monogr 15: 381–385
6. Japanese Society for Esophageal Disease (1976) Guidelines for clinical and pathologic studies on carcinoma of the esophagus. Jpn J Surg 6: 70–86

7. Uematsu S (1982) Roentgenologic diagnosis of depth of invasion of esophageal carcinoma (in Japanese). Rinsho Hoshasen 27: 1037
8. Kumegawa H (1985) Study of endoscopic ultrasonography for esophageal carcinoma (in Japanese). Jpn J Gastroenterol Surg 18: 1174
9. Greenwood M. The "errors of sampling" of the survivorship tables Reports on Public Health and Medical Subjects, no. 33, Appendix 1. In: Cutler SJ (1958) Maximum utilization of the life table method in analyzying survival. J Chron Dis 8: 699–712

Index

MIX
Papier aus verantwortungsvollen Quellen
Paper from responsible sources
FSC® C105338

If you have any concerns about our products,
you can contact us on
ProductSafety@springernature.com

In case Publisher is established outside the EU,
the EU authorized representative is:
Springer Nature Customer Service Center GmbH
Europaplatz 3, 69115 Heidelberg, Germany

Printed by Libri Plureos GmbH
in Hamburg, Germany